Legal Disclaimer

This information contained in this book in both published hard copy and electronic digital PDF file download ("the/this work") is intended for educational purposes only and is not to be used for the actual practice of engineering in any way, shape, or form.

While every effort has been made to achieve a work of high quality, neither the author, contributors, or publishers of this work guarantee the accuracy of or completeness of, or assume any liability in connection with, the information and opinions contained in this work. The author and the publisher shall in no event be liable for any personal injury, property, or other damages of any nature whatsoever whether special, indirect, consequential, or compensatory directly or indirectly resulting from the publication, use of, or reliance upon this work.

In no way does this document represent the views or opinions of the National Council of Examiners for Engineering and Surveying non-profit corporation or any other professional society or organization.

Electrical PE Review, INC is not affiliated with the National Council of Examiners for Engineering and Surveying (NCEES®), the non-profit organization that develops, administers, and scores the Principles and Practice of Engineering (PE) examination. The NCEES® website is located at www.ncees.org.

NCEES® is a registered trademark of the National Council of Examiners for Engineering and Surveying non-profit corporation.

National Electrical Code® and NEC® are registered trademarks of the National Fire Protection Association, INC.

Electrical PE Review® is a registered trademark of Electrical PE Review, INC.

Check for Updates and Corrections

You can verify whether this copy of the practice exam is up to date and check for any corrections or known mistakes by using the camera app on your smartphone to scan the following QR code:

SCAN FOR UPDATES:

PRACTICE EXAM UPDATES

Scan the following QR codes using your smartphone for additional resources:

Email Zach Stone, PE:

More Practice Exams:

Review Course Free Trial:

Live Class - Learn More:

YouTube Channel:

How to Pass the PE Exam:

Table of Contents

About the Author

My name is Zach Stone, PE, the online instructor for Electrical PE Review, INC. Every year, I help hundreds of engineers just like you pass the *NCEES® Electrical Power PE Exam* in order to receive the prestigious title of professional engineer. Here is my 10-second introduction:

- Professionally licensed engineer in the state of Florida.
- ABET-accredited BSEE from a state university.
- Passed both the FE and PE exam on the first try.
- Experienced in the electrical design of commercial buildings, healthcare facilities, and industrial power distribution.
- Experienced in the planning, construction, maintenance, and operation of industrial measurement and instrumentation, process control equipment, automation, motor control centers, utility power generation, medium voltage substations, medium voltage transmission lines, and more.
- I enjoy sharing my understanding of the math and theory behind electrical power engineering to help engineers pass the PE exam.

This practice exam is from our popular online class, located at www.electricalpereview.com, for the new CBT format of the *NCEES® Electrical Power PE Exam*. Our online class includes over 400 videos of worked-out problems, 450 practice problems with full solutions, practice exams, and over 60 hours of live online class instruction.

Each question in this practice exam has been designed and selected to help prepare you for the new CBT format of the *NCEES® Electrical Power PE Exam* and is based on the most up-to-date *NCEES® Exam Specifications*.

Have a question or found an error?

I read and respond to every email I receive. Please feel free to email me directly at zach@electricalpereview.com.

You may also use the camera app on your smartphone to scan this QR code to send me an email ☛

About the New Computer-Based Test (CBT) Format

What Are AIT Practice Problems?

Before the transition to the new computer-based test (CBT) format, the PE exam's previous paper-and-pencil format only included multiple-choice questions. The following problem is a quick example of a multiple-choice question.

Multiple choice:

Which of the following is an example of an even number?

A) 1
B) 2
C) 3
D) 5

Answer: B) 2

With the transition from the paper-and-pencil format to the CBT in December 2020, *NCEES®* introduced four new types of exam questions in addition to multiple choice. These new exam questions are known as **alternative item type (AIT)** questions.

The four new AIT question types are **multiple correct, point and click, drag and drop,** and **fill in the blank.** The following problems are quick examples of each of the new AIT question types.

Multiple Correct:

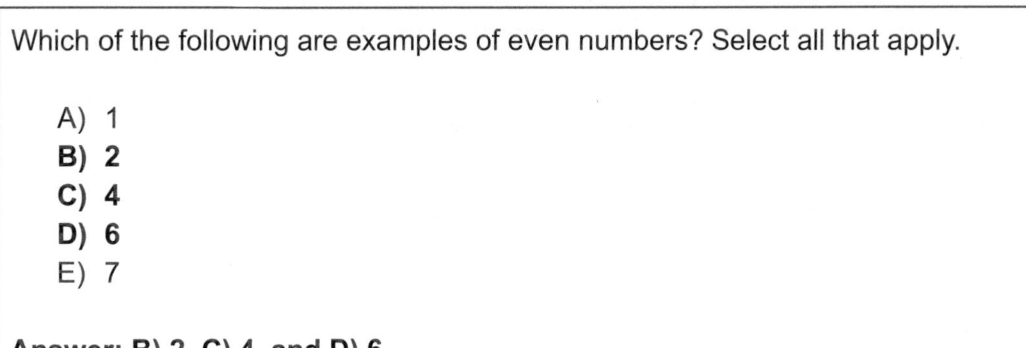

Which of the following are examples of even numbers? Select all that apply.

A) 1
B) 2
C) 4
D) 6
E) 7

Answer: B) 2, C) 4, and D) 6.

Point and Click:

Click the point that is on the circumference of the circle:

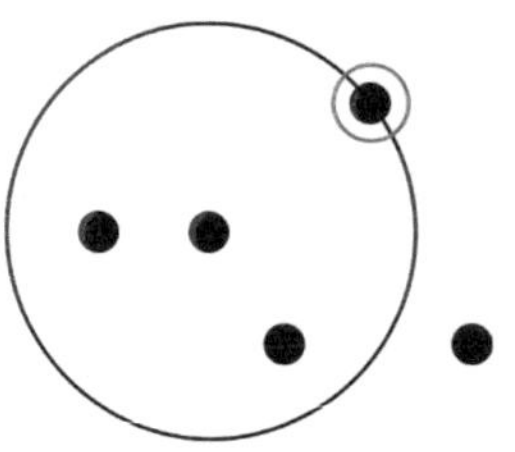

Answer: The correct point is shown circled in red.

Drag and Drop:

Correctly identify the name of each shape by dragging each label to the correct location:

RECTANGLE TRIANGLE CIRCLE SQUARE

Answer: See each arrow point to the correct shape.

Fill in the Blank:

The number 10.579 rounded to the nearest whole number is:___________.

Answer: 11.

What Else is New with the CBT Format?

The largest change to the exam, besides the fact that it is now taken on a computer at a testing center with no outside references besides what NCEES® provides, is that each test taker is given a different randomized version of the PE exam with different questions.

The older paper-and-pencil format circa 2020 was only offered twice a year, once in April and once in October. Although there might have been a different version of the exam each time it was offered, each person taking the PE exam on the same date had the same exact version of the exam.

For example, each test taker in April took the same exact exam, problem for problem, as the others taking the test in April.

This is no longer true for the CBT format. Instead, the exam is offered year-round, and every exam taker will have a completely different randomized version of the exam compared to their peers.

While the number of questions in each exam subject will still fall in the range of the "number of questions" shown on the *NCEES® Exam Specifications*, the actual questions and exact number of questions per exam subject will be randomized and completely unique to each test taker.

It is assumed that even if you take the exam on the same day and at same testing location as a friend or colleague, you both will have completely different randomized version of the Electrical Power PE Exam.

What are Pretest Questions?

Pretest questions are a way for *NCEES®* to evaluate potential questions to include in future versions of the PE exam. It is assumed that *NCEES®* evaluates future potential questions based on how you, the test taker, answered them and based on how you scored on the exam.

Pretest questions are included in your exam when you take the PE exam; however, they are not counted towards your overall score. They will not earn you points for correct answers, and they will not detract from your score for incorrect answers.

While taking the exam, there is no way to tell how many pretest questions you were given, and which of the questions are pretest questions, so answer all exam questions diligently. If you do not pass the exam, it is currently possible to tell the total number of pretest questions based on your exam diagnostic report.

About This Practice Exam

Why Publish an AIT ONLY Practice Exam?

Since the transition to the new CBT format is relatively new for the Electrical Power PE Exam, there is close to nothing in terms of AIT-focused study material on the market to help prepare for the exam, until now.

The vast majority of PE exam questions will be multiple choice with only a handful of AIT questions scattered throughout. However, each AIT practice problem that you are able to correctly master with the help of this book represents an increase in MANY possible correct answers of the exam's multiple-choice questions, which tend to be much easier to solve.

This practice exam follows the *NCEES® Exam Specifications* to meet the range of the number of questions in each subject; however, every single question in this practice exam is an AIT question.

What Are the Benefits of Studying AIT Practice Problems?

Compared to multiple-choice questions, AIT questions can be much more difficult. Depending on the problem type, such as multiple correct, there can be more than one selection required for the problem to be correct with no partial credit for a partially correct answer.

Because of this, you can think of studying AIT PE exam practice problems as learning how to solve multiple problems in one. This will require you to really challenge and refine your understanding of each PE exam subject area, especially when it comes to maximum or minimum values, interpreting graphs, proper rounding, and how each variable in any given relationship relates with the others, such as torque vs speed for induction motors.

Each AIT practice problem you are able to solve correctly represents a relationship that you fully understand for the PE exam.

Each AIT practice problem that you are able to understand after working through the solution represents a new relationship that you now have a much stronger grasp of, and that can be applied to a higher number of exam questions, especially the much more common plain old multiple-choice-type question.

What Are the Goals and Intentions of This Book?

The primary goal of this book is to help you *dramatically* improve your PE exam score by increasing your understanding across all exam subjects as quickly as possible.

It is intended to be a powerful learning tool that will rapidly improve your overall exam proficiency. The average level of difficulty for each problem will be greater than what you can expect to see on the PE exam. This is intentional. You will be challenged. You will be pushed. You will learn far more than you ever thought possible for the Electrical Power PE Exam from one single practice exam.

You will also enjoy how much your other practice exam scores improve after taking the time to fully internalize the lessons contained in every single solution to each problem in this practice exam.

For benchmarking purposes, a strong first- or second-time score for taking this practice exam is 50% to 60%.

How to Use This Practice Exam

Depending on how far along you are in your preparation for the PE exam, there are two different ways you can use this practice exam.

The first, if you are just starting to study, is to use this practice exam as just a large volume of practice problems to study and learn from.

The second is to treat this like the real PE exam by completing it in an eight-hour sitting with a one-hour break in between the morning and afternoon session. This is the preferred way to use this practice exam since it will help get you accustomed to solving problems you are unfamiliar with, help you become more familiar with using the *NCEES® Reference Handbook* and Code Books, and prepare you for the pace required to work problems for the extended duration of the eight-hour PE exam.

To take this practice exam like the real PE exam, consider the following approach:

- Go somewhere quiet and distraction-free.
- Bring the *NCEES® Reference Handbook* and Code Books in either digital or hard copy format.
- Set a four-hour timer for the first 40 problems and do not stop to check the solutions or to see check which problems you answered correctly. Take a 45-minute to one-hour lunch break, then set the timer for another four hours and complete the remaining 40 problems. Once you have completed the first 40 problems, do not go back and only work on the remaining 40.
- Instead of answering each question in the order that it appears, complete each session in multiple passes by answering the easiest questions first, the medium difficulty questions second, and the hardest questions last. Taking the exam in multiple passes is one of the most successful test-taking strategies for the PE exam.
- On your first pass, you should only be answering questions you are comfortable solving quickly and without help from references.
- On your second pass, focus on answering all leftover questions with the help of the *NCEES® Reference Handbook* and Code Books. However, if you find yourself getting lost in references and wasting time, you should most likely skip the current question and save it for the third pass.

- The third pass should be all remaining questions. However, if you find yourself at a standstill or just unsure of how to answer it, skip it and keep moving forward.

- The last 15 to 30 minutes on your four-hour timer should be spent trying to answer the last handful of the more difficult questions.

- Grade your exam only after you have completed the entire exam over an 8-hour time period. On a separate piece of paper, make a list of every question you got incorrect, including questions you had to make an educated guess on, even if you got it correct. Include the date that you took the sample exam.

- Over the course of the following few weeks, work through the solutions of the questions that you got wrong and the questions you guessed on. Seek out more questions in each of those subjects (the quiz questions in our online review course are a great place to start).

- Several weeks after you've completed working through the solutions, try attempting the same set of questions you made a list of that you got wrong (or that you had to guess on) during the first complete attempt of this practice exam. When you are able to answer each of these questions correctly, cross it off the list. Consider doing this every few weeks to narrow down the list, to learn from repetition, and to keep challenging yourself with the problems you find difficult so that you retain the material.

Answer Sheet

Name:______________________________ Date:____________________

Attempt #:_______ # of Correct Questions:_____ Percent Score out of 80:___________

1.	A B C D E
2.	
3.	See Drag and Drop Diagram
4.	
5.	
6.	A B C D E
7.	
8.	A B C D E
9.	
10.	
11.	A B C D E
12.	A B C D E
13.	
14.	A B C D E
15.	
16.	See Drag and Drop Diagram
17.	A B C D E
18.	
19.	
20.	See Drag and Drop Diagram
21.	
22.	A B C D E
23.	
24.	
25.	
26.	
27.	A B C D E
28.	A B C D E
29.	A B C D E
30.	See Point and Click Diagram
31.	See Point and Click Diagram
32.	
33.	
34.	See Drag and Drop Diagram
35.	
36.	See Drag and Drop Diagram
37.	See Drag and Drop Diagram
38.	A B C D E
39.	
40.	A B C D E
41.	See Drag and Drop Diagram
42.	A B C D E
43.	A B C D E
44.	
45.	A B C D E
46.	
47.	A B C D E
48.	A B C D E
49.	A B C D E
50.	A B C D E
51.	See Drag and Drop Diagram
52.	A B C D E
53.	A B C D E
54.	
55.	
56.	A B C D E
57.	
58.	A B C D E
59.	See Drag and Drop Diagram
60.	See Drag and Drop Diagram
61.	A B C D E
62.	A B C D E
63.	See Drag and Drop Diagram
64.	A B C D E
65.	A B C D E
66.	A B C D E
67.	See Drag and Drop Diagram
68.	
69.	A B C D E
70.	See Drag and Drop Diagram
71.	A B C D E
72.	A B C D E
73.	A B C D E
74.	A B C D E
75.	A B C D E
76.	A B C D E
77.	A B C D E
78.	A B C D E
79.	
80.	See Drag and Drop Diagram

Diagnostic Report

If you don't pass the PE exam, NCEES® will send you a diagnostic report, similar to the *Scoring Evaluation Table* shown below, that demonstrates how well you performed in each subject.

After you grade your *Answer Sheet* from the previous page, use the *Answer Key* located in the back of this practice exam to fill in the **Number of Questions You Answered Correctly per Subject** column in the table below.

Once you've done this for each of the nine major subjects, fill in the **Percent of Questions Answered Correctly per Subject** column by dividing the number of correctly answered questions per subject by the total number of questions per subject, multiplying by 100 to convert the decimal to a percentage.

Use this information to help determine where you should be spending your time studying in order to help improve your overall score.

Scoring Evaluation Table:

Exam Subject	**Number of Questions You Answered Correctly per Subject**	**Total Number of Questions in the Practice Exam per Subject**	**Percent of Questions Answered Correctly per Subject**
Ch. 1 - Measurement and Instrumentation		6	
Ch. 2 - Applications		9	
Ch. 3 - Codes and Standards		11	
Ch. 4 - Analysis		10	
Ch. 5 - Devices and Power Electronic Circuits		6	
Ch. 6 - Induction and Synchronous Machines		7	
Ch. 7 - Electric Power Devices		9	
Ch. 8 - Power System Analysis		10	
Ch. 9 - Protection		12	

1. Select the two-wattmeter method connection diagram below that will result in accurate power measurement for a three-phase, three-wire, unbalanced system. Select all that apply.

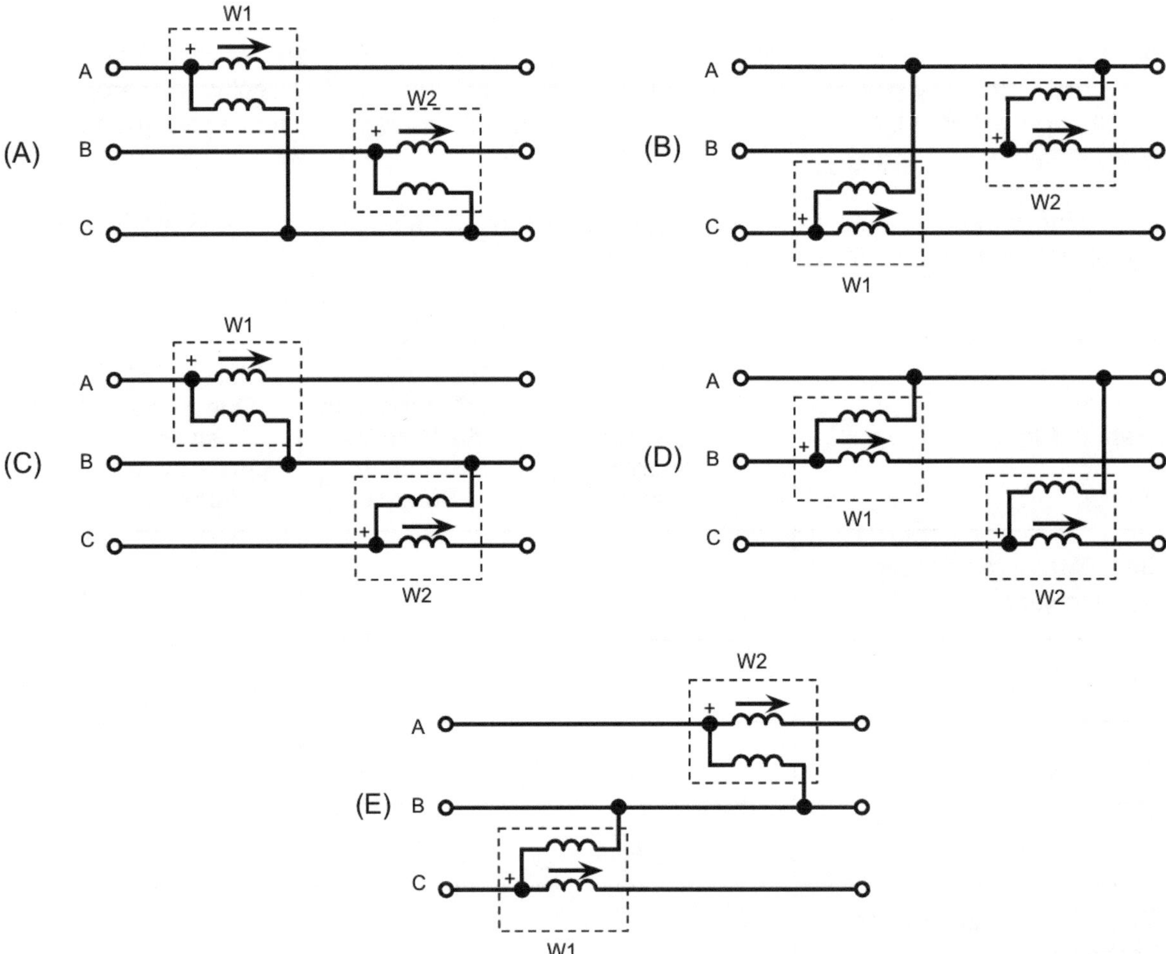

2. The 69 kV distribution system shown below provides power to a 55 MVA, 69-15 kV three-phase transformer. At 100% transformer load, the current signal input to the ANSI #51 relay to the nearest ampere is _________ A

Fill in your response.

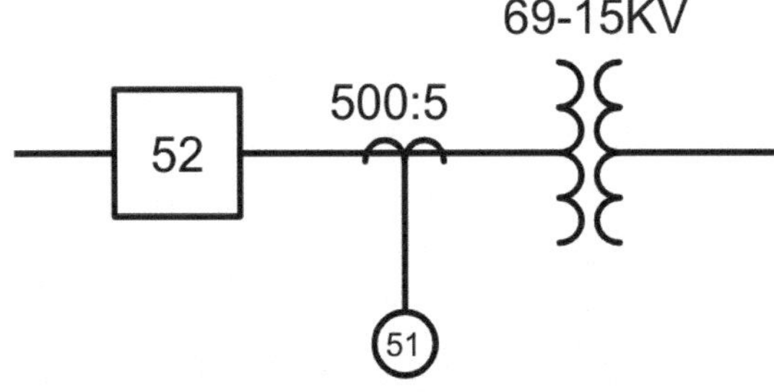

3. Shown below are the general characteristics of the different insulation currents during an insulation test for a generally healthy cable.

   Drag the names of each of the individual insulation currents to their correct curve on the microamp vs linear time scale graph.

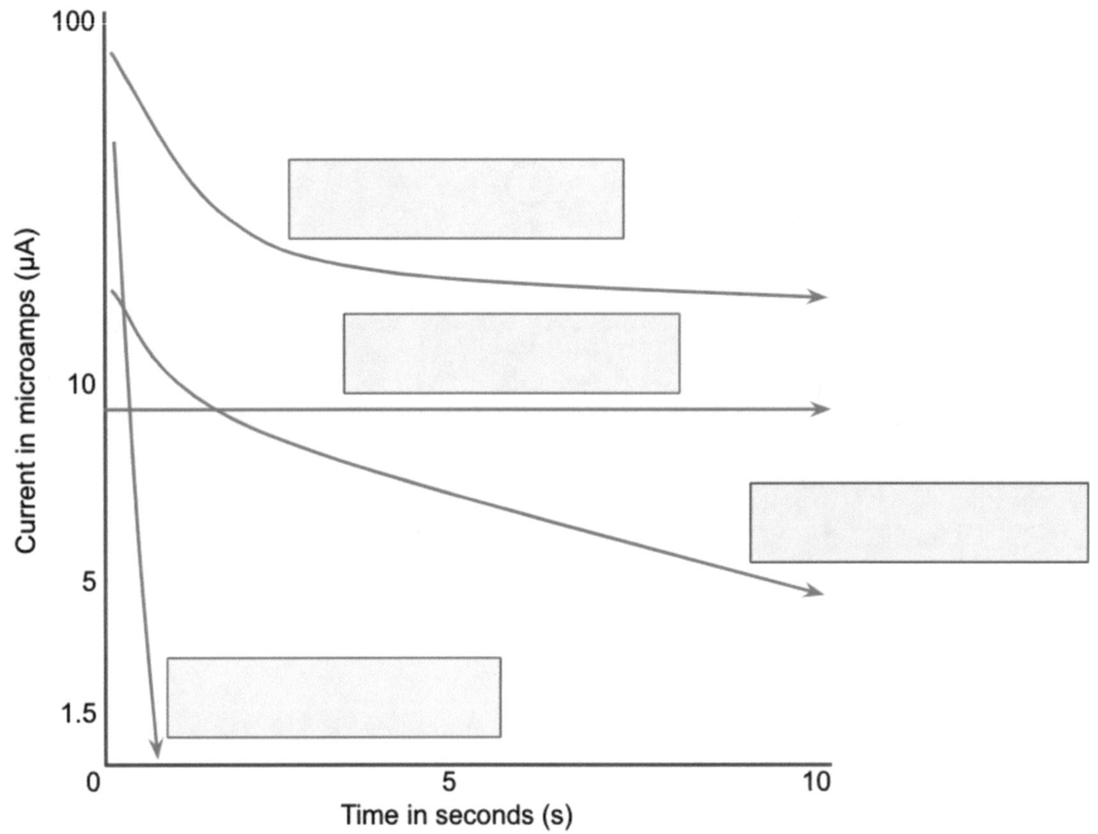

Dielectric Absorption Current	Total Insulation Current	Leakage Conduction Current	Capacitive Charging Current

4. A 2 kV rated conductor with thermoset insulation has 125 megaohms of insulation resistance when measured using a 1 kV DC test voltage at 65 degrees Celsius.

   The insulation resistance corrected to 40 degrees Celsius to the nearest unit shown is _________ MΩ.

   Fill in your response.

5. Four grounding rods are used to measure the resistance to ground at a location with a soil resistivity of 25,000 ohm centimeters. The outer two rods are 300 feet apart while the inner two rods are 50 feet apart. Use the simplified calculation in which a>>2b.

   The calculated resistance to ground rounded to one decimal place is _________ Ω

   Fill in your response.

6. What are the advantages that are unique to the driven rod method compared to other ground resistance testing methods? Select all that apply.

(A) The ability to take a ground resistance measurement at the testing location.

(B) Only three rods are required for the test compared to four.

(C) The ability to determine how far permanent ground rods will be able to be driven in the testing location.

(D) Greater accuracy in ground measurement compared to non-driven test methods.

(E) All three testing rods may be equal in length compared to other methods.

7. A rectangular building 75 feet long, 50 feet wide, and 30 feet high is located in an area that is surrounded by trees much taller than the building.

For the purpose of lightning protection, the collection area of the building rounded to the nearest unit shown is _________ ft^2

Fill in your response.

8. Shown below is a reliability block diagram of the assets that make up a crucial system inside a manufacturing plant.

   Select all of the following statements that are true.

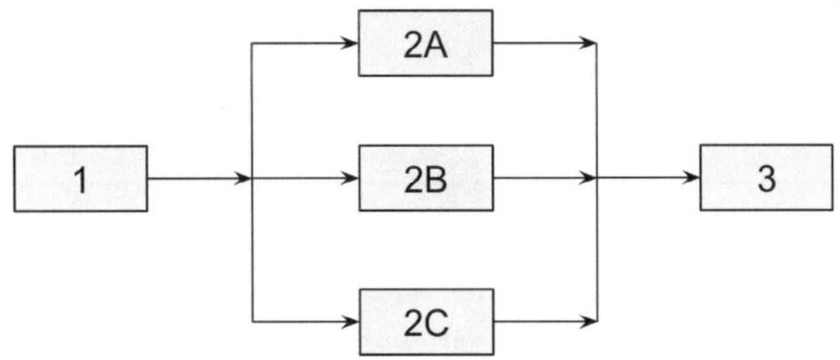

(A) In order for the system to be operable, all five assets in the reliability block diagram must also be operable.

(B) The system will be operable if assets 1, 2A, and 3 are operable.

(C) The system will be operable if assets 1, 2B, and 3 are operable.

(D) The system will be operable if assets 1, 2B, 2C, and 3 are operable.

(E) The system will be operable if assets 1 and 3 are operable.

9. The minimum mounting height for long distribution roadway lighting with a luminous intensity of 30,200 candela to the nearest foot is _________ ft.

   Fill in your response.

10. The table below shows the maximum demand data for separate area processes of a local wastewater treatment plant that has an overall average maximum billing demand of 6.9 MW each month.

	Megawatts	Peak Time
Incoming water	5	4 PM
Water processing	7.5	1 PM
Water holding and storage	2.3	3 PM

The percent coincidence factor for the wastewater treatment plant rounded to one decimal place is _________ %.

Fill in your response.

11. Neglecting any possible utility-imposed power factor penalty, which of the following statements are true for the energy cost associated with operating an induction motor? Select all that apply.

(A) The amount of real power drawn by the motor from the power system is paid for at the energy cost rate.

(B) The amount of apparent power drawn by the motor from the power system is paid for at the energy cost rate.

(C) The amount of real power delivered by the motor to the load is paid for at the energy cost rate.

(D) Both the amount of real power delivered by the motor to the load and the total real power loss are paid for at the energy cost rate.

(E) All three are paid for at the energy cost rate: the amount of real power drawn by the motor from the power system, the amount of real power delivered by the motor to the load, and the total real power loss.

12. A lending institution charges 1% interest per month. Which of the following statements are true? Select all that apply.

    (A) The effective interest rate per month is 1%

    (B) The nominal monthly interest rate is 1%

    (C) The nominal annual interest rate is 1%

    (D) The nominal annual interest rate is 12%

    (E) The effective annual interest rate is 12.7%

13. An oil company purchased new drilling equipment for $55,000 that is legally classified as a 5-year property class asset.

    The final depreciation deduction for the asset according to the modified accelerated cost recovery system is $_________.

    Fill in your response.

14. In most applications, which power system grounding technique out of the choices below will allow the system to stay energized even in the presence of a single line-to-ground fault? Select all that apply.

(A) Solidly grounded wye power source

(B) High impedance grounded wye power source

(C) Ungrounded wye power source

(D) Ungrounded delta power source

15. Gravel with a resistivity of 3 kiloohm meters is added to the surface of a substation 0.1 meters deep to increase the contact resistance of the soil that has a resistivity of 150 ohm meters.

Using a foot radius of 0.08 meters, the surface layer derating factor used in step voltage calculations rounded to one decimal place is _________.

Fill in your response.

16. The line current phasor diagrams for four different three-phase systems are shown below. Match the correct label for each system by dragging each label to its appropriate phasor diagram.

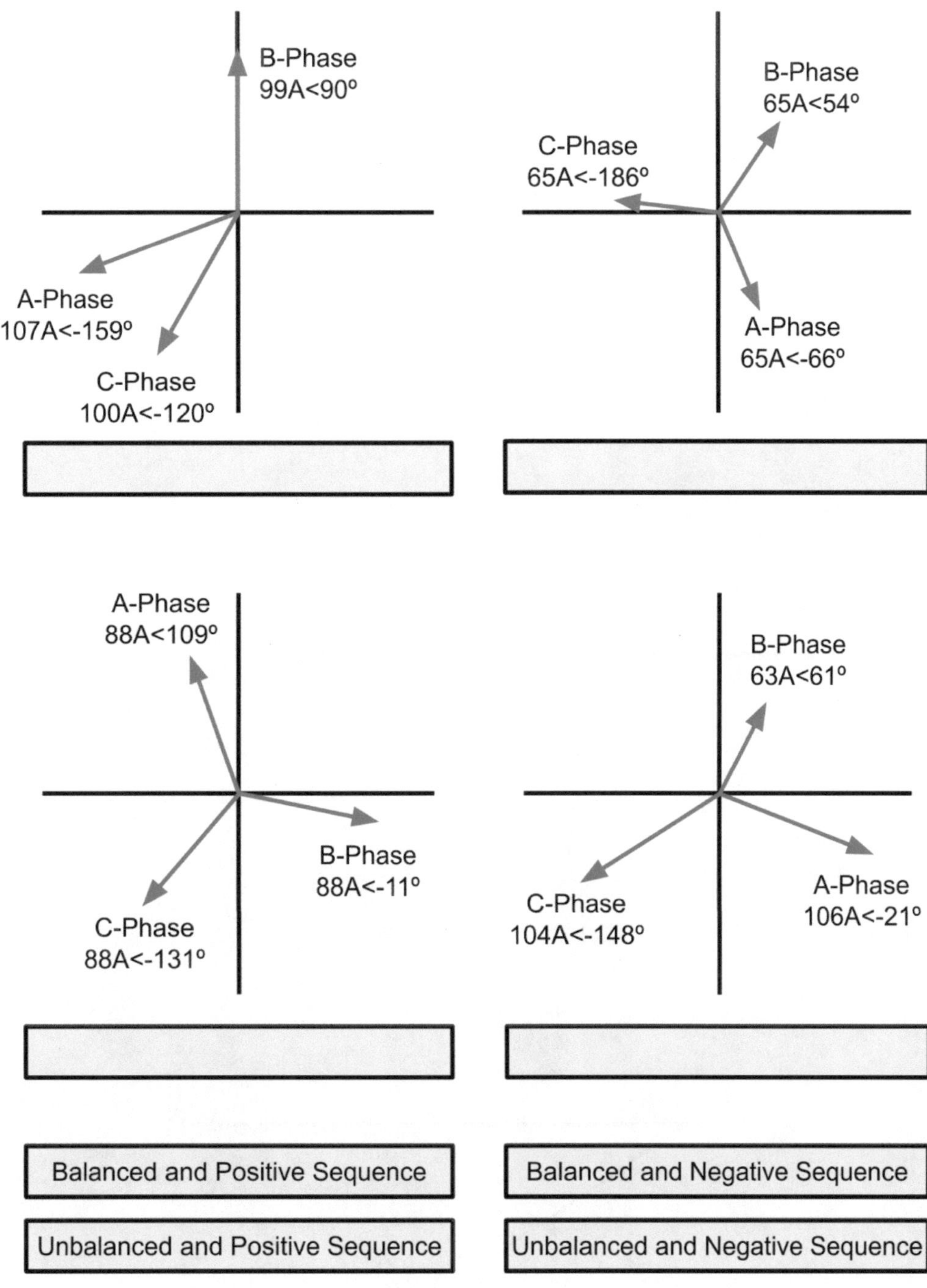

17. Which of the following voltage phasor diagrams represent a lagging synchronous machine? Select all that apply.

E_o = Internal phase voltage.
E = Terminal phase voltage.

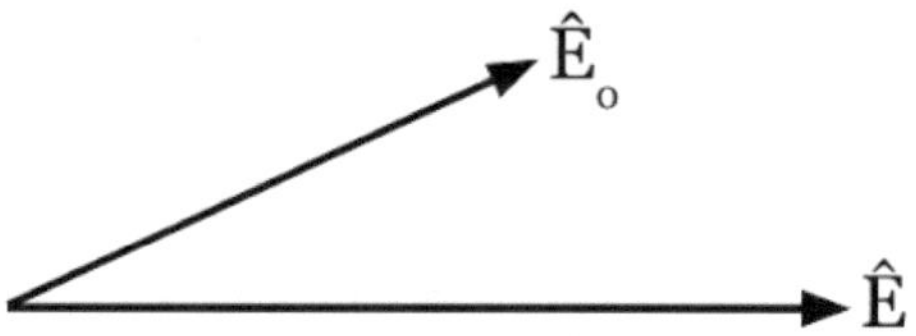

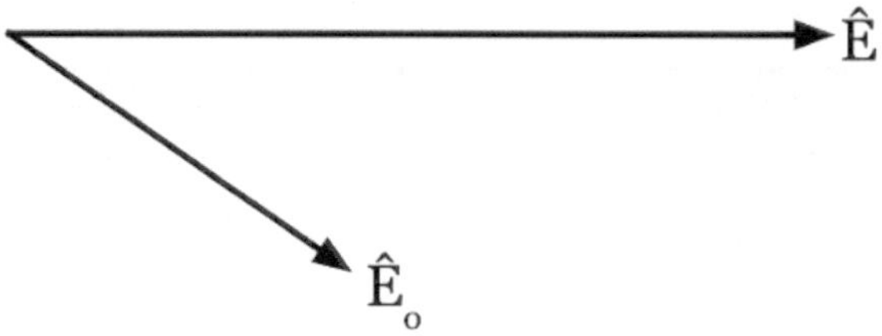

A

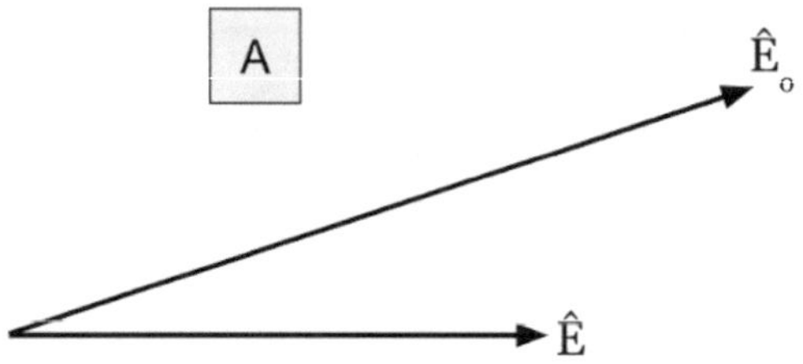

$\hat{E}_o$

$\hat{E}$

B

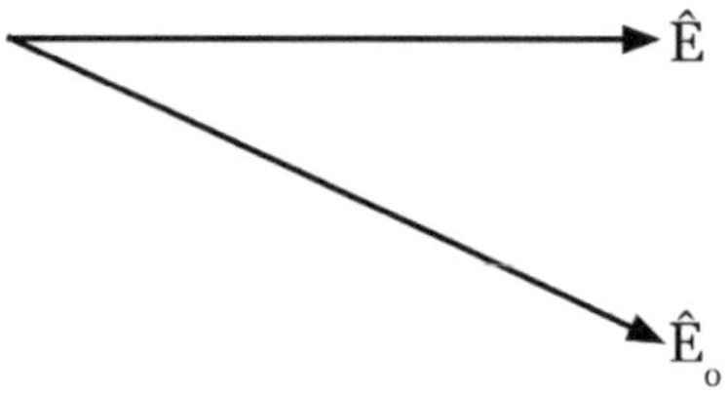

$\hat{E}$

$\hat{E}_o$

C

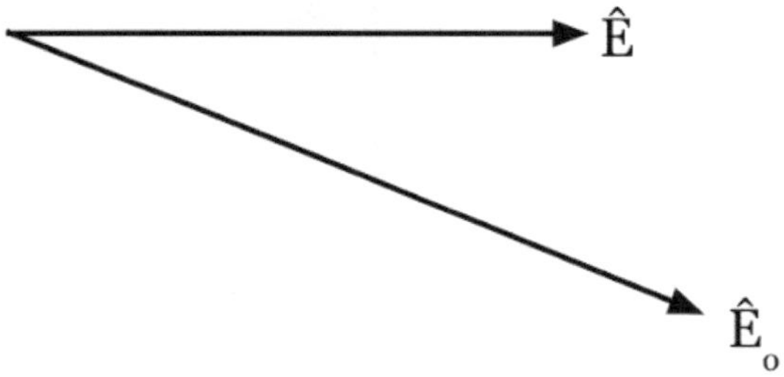

$\hat{E}$

$\hat{E}_o$

D

$\hat{E}_o$

$\hat{E}$

E

$\hat{E}$

$\hat{E}_o$

18. The magnitude of the line current delivered by each phase of the balanced and positive sequence, three-phase, 60 Hz, 15 kV power supply shown below rounded to the first decimal place is _________ A.

    The line impedance in each phase (R + jX) is 2 + j8 ohms.
    The delta connected load impedance in each phase (Z) is 3,600 + j65 ohms.

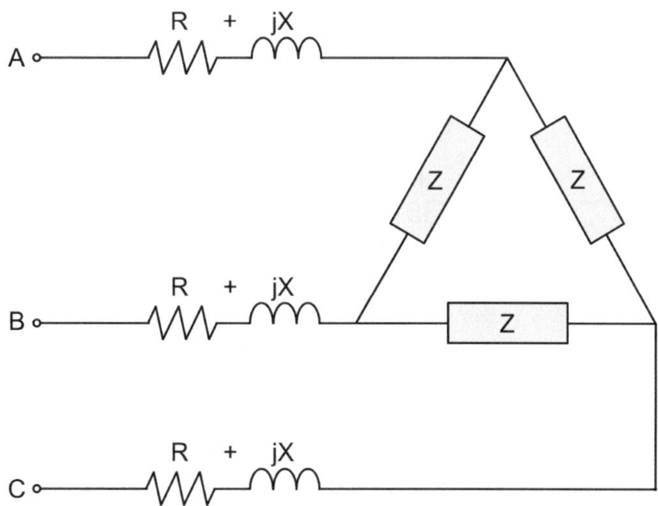

Fill in your response.

19. An unbalanced delta connected load is powered by a three-phase, 208 volt power supply:

    A-phase: 1 + j6 ohms
    B-phase: 2 + j2 ohms
    C-phase: 5 + j ohms

The magnitude of the balanced wye equivalent load rounded to one decimal place is _________ Ω.

Fill in your response.

20. Below are four different types of circuit diagrams that belong to the same system. Match the correct label for each circuit diagram by dragging each label to its appropriate choice.

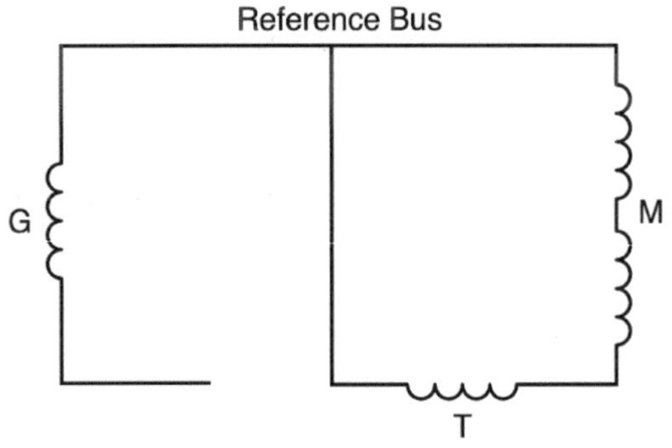

Reference Bus
G M T

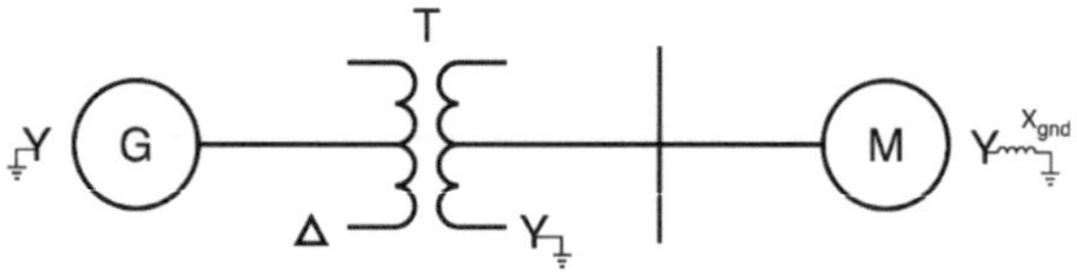

T
Y G Δ Y M Y X_{gnd}

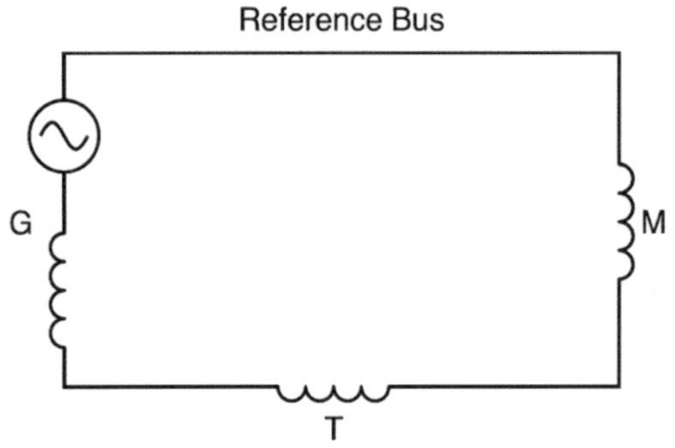

Reference Bus
G M T

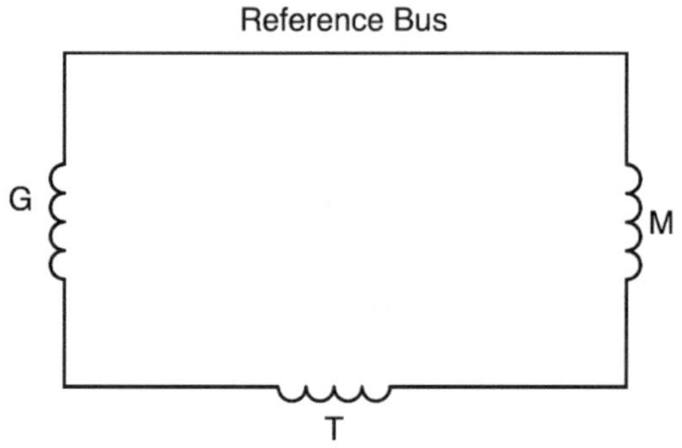

Reference Bus
G M T

Negative Sequence Network	Positive Sequence Network
Zero Sequence Network	Single Line Diagram

21. A single line-to-ground fault occurs on bus A in the three-phase system shown below. The total equivalent complex impedance of the zero sequence network rounded to two decimal places is _________ + j _________ pu.

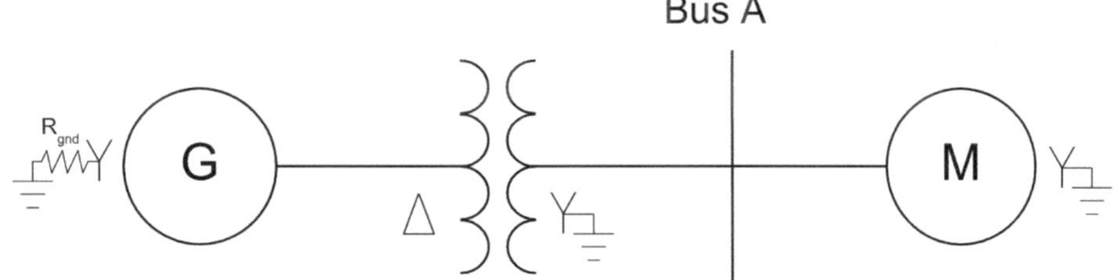

Assume all per unit values are in the same base:

$X_G^{(0)} = 0.45$ pu
$R_{gnd} = 1.9$ pu
$X_T^{(0)} = 0.67$ pu
$X_M^{(0)} = 0.80$ pu

Fill in your response.

22. When working with the per unit system, which of the following is true? Select all that apply.

(A) The system base power and base voltage are the same in every voltage zone.

(B) The system base power is the same in every voltage zone; the voltage base changes depending on each transformer ratio.

(C) The system base impedance and base voltage in each voltage zone change depending on each transformer ratio.

(D) When the voltage is equal to the base voltage, the short circuit per unit fault duty is equal to the short circuit per unit current.

(E) When the voltage is equal to the base voltage, the short circuit per unit current is inversely proportional to the total equivalent per unit impedance.

23. Assuming that the switch in the circuit below is turned on (closed) at t = 0 seconds and there is no initial voltage across the capacitor, the capacitor will be charged to steady state voltage conditions at a time of _________ milliseconds. Round to the nearest millisecond.

    V = 24 VDC, R = 9.5 kΩ, C = 6 μF

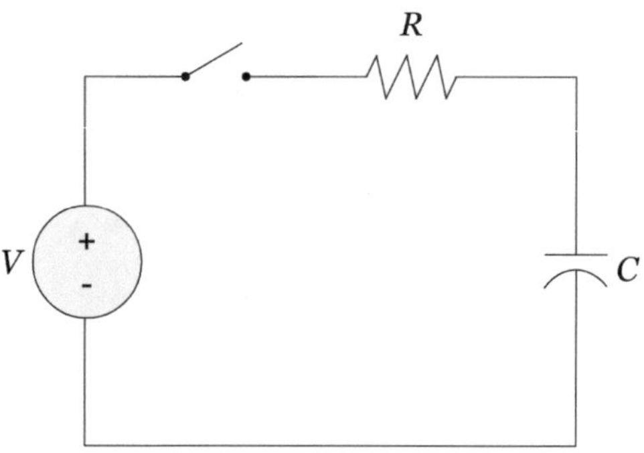

    Fill in your response.

24. The switch in the circuit below is turned on (closed) at t = 0 seconds. Prior to the switch closing, 14 volts DC was measured across the capacitor. The current flowing through the capacitor at t = 60 milliseconds is _________ milliamperes. Round to one decimal place.

    V = 35 VDC, R = 7.5 kΩ, C = 4 μF

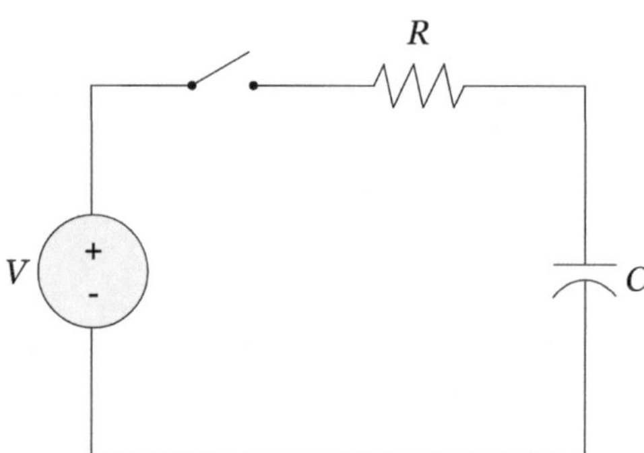

    Fill in your response.

25. Assuming that the current flowing through the inductor prior to the switch closing in the circuit below is zero, the voltage across the inductor at the exact moment that the switch closes is _________ volts. Round to the nearest volt.

    V = 45 VDC, R = 1 kΩ, L = 2 H

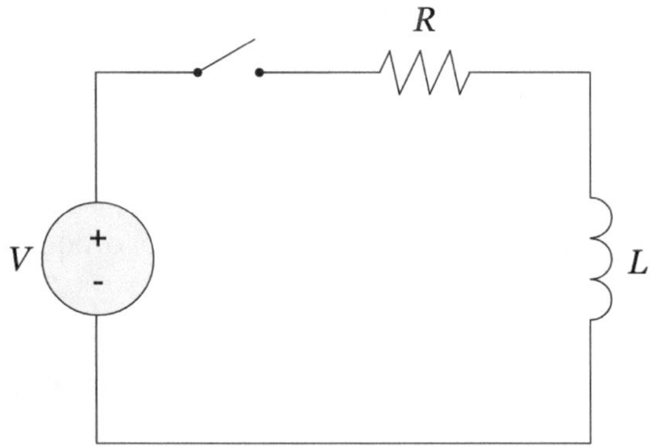

Fill in your response.

26. A secondary rechargeable lithium-ion battery is being tested for use in a new consumer electronic device that is being developed. The voltage potential that can be expected daily from each lithium-ion cell when powering the electronic device is most likely to be _________ volts. Round your answer to one decimal place.

    Fill in your response.

27. Which of the following statements regarding battery behavior and characteristics are false? Select all that apply.

    (A) In general, an increase in current delivered by a battery will result in a decrease in the total amount of amp hours the battery is able to deliver.

    (B) In general, a battery in an environment with an ambient temperature higher than room temperature will typically result in an increase in the total amount of amp hours the battery is able to deliver.

    (C) A battery in an environment with an ambient temperature lower than room temperature will result in an increase in the total amount of amp hours the battery is able to deliver.

    (D) A battery will supply more DC current in hotter temperatures at a constant amp hour capacity.

    (E) A battery will supply more amp hours in hotter temperatures at a constant discharge current.

    (F) A battery can only be expected to supply 100% rated capacity at higher than room temperature.

28. A capacitor is placed in parallel with a DC load fed by a single-phase half-wave uncontrolled rectifier. Select all statements below that are true.

    (A) The peak to peak ripple voltage is inversely proportional to the farad rating of the capacitor.

    (B) The peak to peak ripple voltage as a percentage of the maximum output voltage will decrease proportionally as the DC load resistance increases.

    (C) The time constant of the circuit is directly proportional to the peak to peak ripple voltage.

    (D) As the capacitance rating of the capacitor increases, the minimum output voltage increases.

    (E) Diode current decreases as the capacitance rating of the capacitor increases.

29. Which of the following duty ratio values can be used for a buck-boost converter to operate as a step up DC to DC converter? Select all that apply.

(A) 0.2

(B) 0.4

(C) 0.6

(D) 0.8

(E) 1.0

30. Click on the correct combination of switches in the full-bridge inverter circuit that are closed during the shaded region of the DC voltage graph shown.

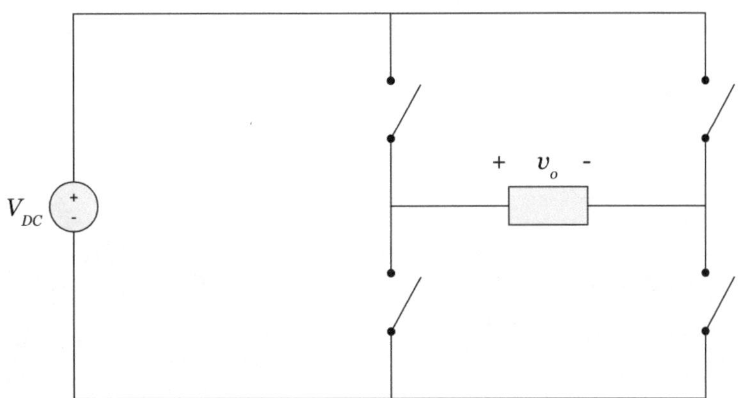

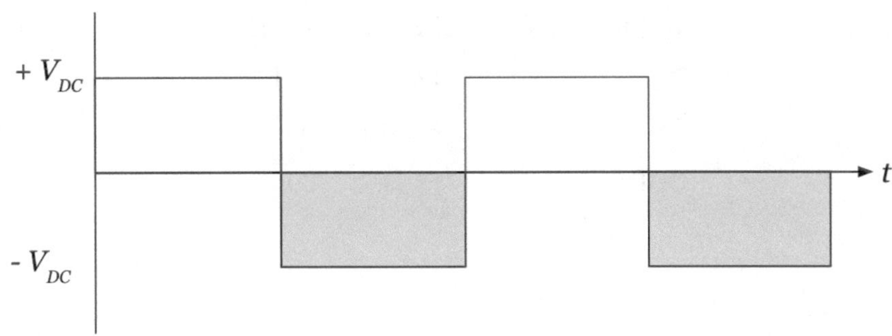

31. A VFD is used to reduce the speed of an induction motor by 50%. Click on the portion of the graph below between the two points that most likely represents the applied voltage (V) and motor current (A) behavior if the VFD applies rated motor voltage.

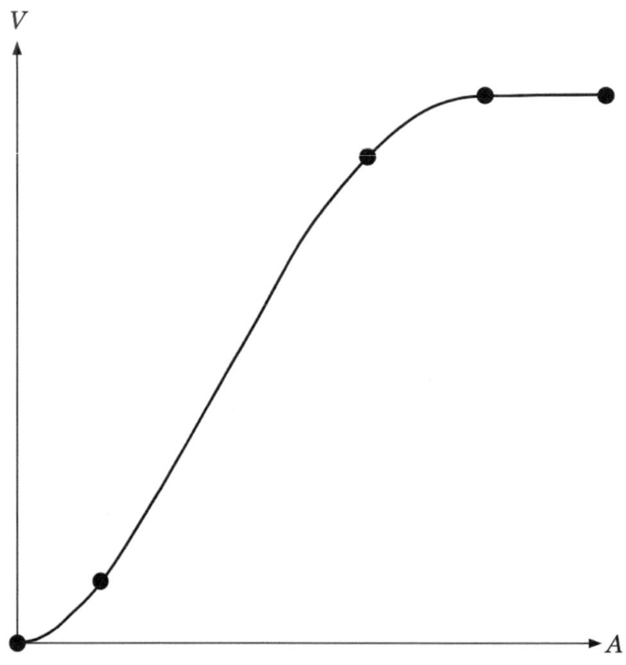

32. The total torque of the motor using the nameplate provided below rounded to the nearest newton meter is _________ N•m.

HP	100	**RPM**	3570
Frame	405TS	**EFF**	95.4
Voltage	208	**Phase**	Three
PF	0.90	**FLA**	241.1
SF	1.15	**NEMA**	B
CODE	G	**Freq**	60

Fill in your response.

33. A three-phase, 25 MVA, 13.8 kV rated salient pole synchronous generator has a 5.12 ohm direct axis reactance and a 3.39 ohm quadrature axis reactance. The electrical power delivered by the machine when the induced phase voltage is 12 kV with an electrical torque angle of 15 degrees is _________ MW.

    Round your answer to the nearest whole unit.

    Fill in your response.

34. Drag and drop the label for each type of power consumed by an induction motor and the leakage reactances to their appropriate circuit elements shown below.

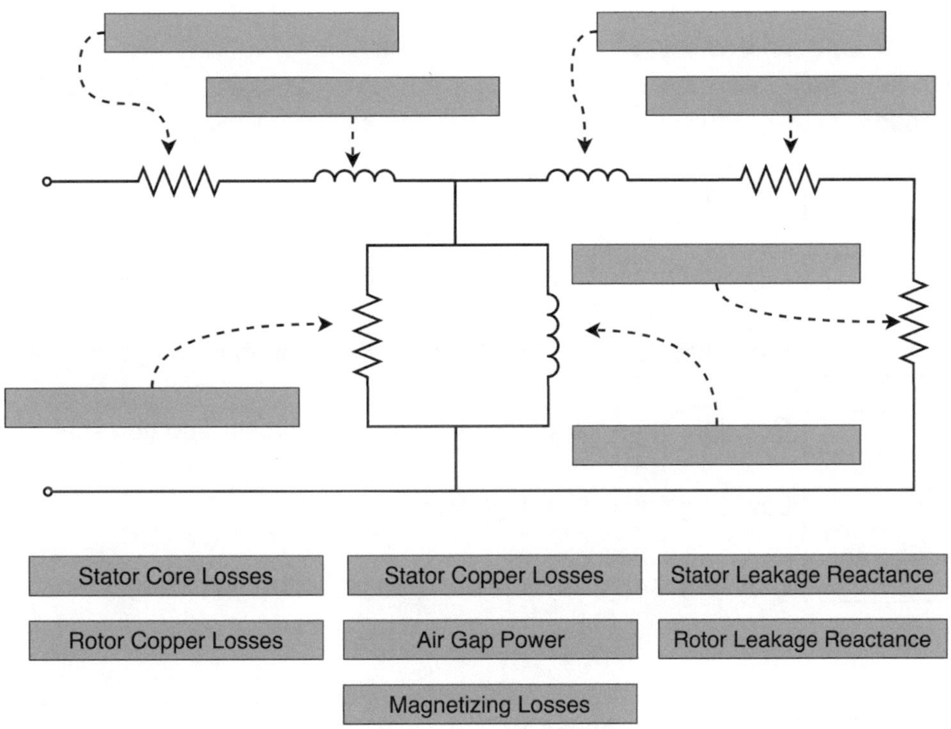

35. Consider the three-phase, 208 V, 60 Hz, wound-rotor induction motor with the following equivalent circuit components:

R_1 =1.45 Ω R_2 = 1.32 Ω X_1 = 2.5 Ω X_2 = 2.8 Ω
R_C = 875 Ω X_M = 112 Ω

The Thévenin-equivalent stator resistance back to the power supply to the nearest ohm is _________ Ω.

Fill in your response.

36. Match the appropriate *National Electrical Manufacturers Association®* (NEMA) motor design letter with their most appropriate application by dragging each label to the correct description.

[ ] High efficiency applications.

[ ] High starting torque applications with standard slip ranges.

[ ] High starting torque applications with a need for large values of slip.

[ ] Constant speed applications with reduced starting current.

[ ] Low starting current applications.

[ ] Constant speed applications with high starting current.

NEMA A	NEMA B	NEMA C	NEMA D	NEMA E	NEMA F

37. Drag the label of each description to the appropriate place on the induction machine torque vs speed curve shown below.

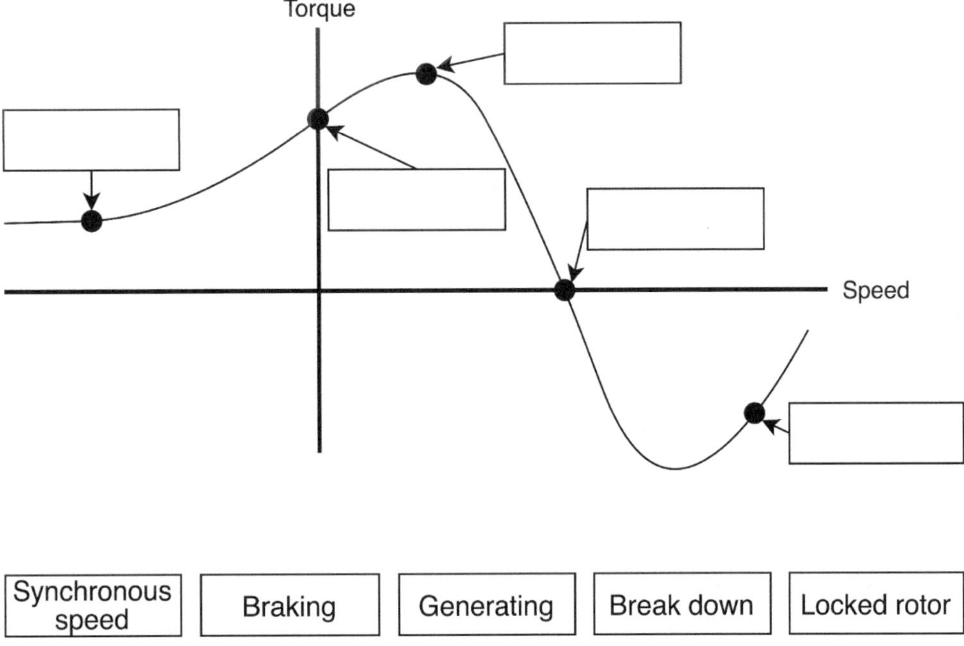

Synchronous speed	Braking	Generating	Break down	Locked rotor

38. Select the motor starting methods below that may be used to limit the starting current of an induction motor. Select all that apply.

(A) Across the line starting

(B) Reduced voltage auto-transformer starting

(C) Reduced voltage resistance starting

(D) Wye start delta run

(E) Soft start speed drive

39. Three single-phase transformers are connected in parallel to provide power to a 480 volt rated load. The current supplied to the shared load by transformer T1 is 189 amps.

Transformer	KVA Rating	Primary Voltage Rating	Secondary Voltage Rating	Percent Impedance
T1	175	2400	480	6.2
T2	150	2400	480	4.5
T3	225	2400	480	5.7

The total current drawn by the shared load to the nearest amp is _________ A.

Fill in your response.

40. Below are statements regarding transformer efficiency, power ratings, and power losses. Select all choices that are true.

(A) The apparent power output of a transformer operating at maximum efficiency can be determined by multiplying the apparent power rating of the transformer by the square root of the ratio of the magnetizing core losses to the full load copper losses.

(B) Maximum efficiency of a transformer occurs when the transformer load is equal to the transformer rated power.

(C) Maximum efficiency of a transformer occurs when the transformer load results in an equal amount of I^2R losses in the transformer compared to the no load losses.

(D) Transformer copper losses are directly proportional to changes in percent load.

(E) As long as the percent load is constant, changes to the load power factor will not have an effect on transformer efficiency.

41. Rank the cooling methods listed below for liquid-filled transformers from the method that results in the highest power output capacity to the method that results in the lowest power output capacity by dragging each label to their correct location.

[ ]	4. Highest power output capacity
[ ]	3. Second highest power output capacity
[ ]	2. Second lowest power output capacity
[ ]	1. Lowest power output capacity

OFAF	ODAF	ONAF	ONAN

42. Out of the choices below, select all of the appropriate applications for reactors used in power systems.

(A) Transient current prevention.

(B) Series transmission line compensation.

(C) Voltage support to increase load voltage.

(D) Harmonic reduction.

(E) Prevent damage to motors fed from variable-frequency drives.

43. The transformer short circuit test can be used to determine which of the following? Select all that apply.

(A) Voltage regulation

(B) Winding impedance

(C) Percent impedance

(D) Core loss resistance

(E) Magnetizing reactance

44. A single-phase, 112 kVA, 15k-600V transformer is shorted on the low-voltage side during a short circuit test with all measurements taken on the high voltage side.

Short Circuit Voltage	500 volts
Short Circuit Current	7.1 amps
Short Circuit Power	1,230 watts

The percent impedance of the transformer rounded to one decimal point is ________ %.

Fill in your response.

45. The transformer open circuit test can be used to determine which of the following? Select all that apply.

(A) Voltage regulation

(B) Winding impedance

(C) Percent impedance

(D) Core loss resistance

(E) Magnetizing reactance

46. Voltage is applied to the high voltage side of a single-phase, 112 kVA, 15k-600V transformer with the low-voltage side left open. All measurements are taken from the high voltage side.

Open Circuit Voltage	15,000 volts
Open Circuit Current	0.15 amps
Open Circuit Power	1,023 watts

The magnetizing reactance of the transformer rounded to the nearest kiloohm is _________ kΩ.

Fill in your response.

47. Out of the choices below, select all of the appropriate applications for capacitors used in power systems.

(A) Power factor correction

(B) Voltage support

(C) Transmission line series compensation

(D) Harmonic filter

(E) Ground fault protection

48. Select the statements below that are true if n equal conductors are connected in parallel for each phase to supply power to a load.

(A) The voltage drop across each phase will increase by a factor of n.

(B) The voltage drop across each phase will decrease by a factor of n.

(C) The current flowing through each conductor will increase by a factor of n.

(D) The current flowing through each conductor will decrease by a factor of n.

(E) The equivalent line impedance for each phase will decrease by a factor of n.

49. Select each statement below that would improve the voltage regulation of a short transmission line with rated voltage at the sending side and a constant load current. Select all that apply.

    (A) Decrease the series line impedance.

    (B) Decrease the shunt line capacitance.

    (C) Improve the load power factor.

    (D) Install a series reactive line compensator.

    (E) Install a recloser.

50. Select each statement below that is true for a three-phase power factor correction wye connected capacitor bank that is reassembled into a delta connected capacitor bank. Select all that apply.

    (A) The voltage applied across each phase of the capacitor bank will increase by a factor of the square root of three.

    (B) The reactive power supplied by the capacitor bank will increase by a factor of three.

    (C) The line charging current drawn by the capacitor bank will increase by a factor of three.

    (D) The reactance of each phase of the capacitor bank will increase by a factor of three.

    (E) The capacitance of each phase of the capacitor bank will increase by a factor of three.

51. Drag and drop the label for each type of electrical fault to the appropriate three-phase voltage waveform.

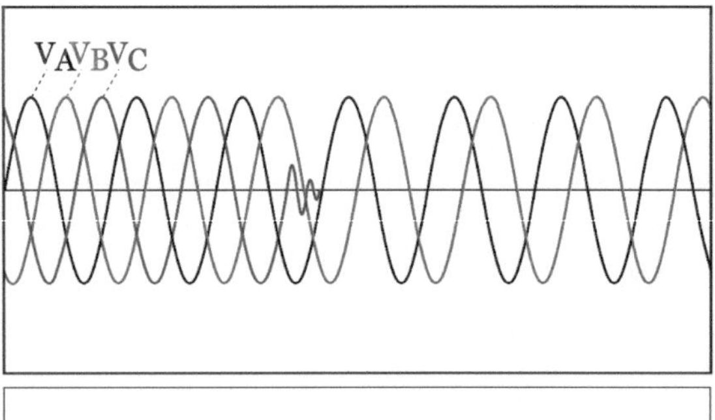

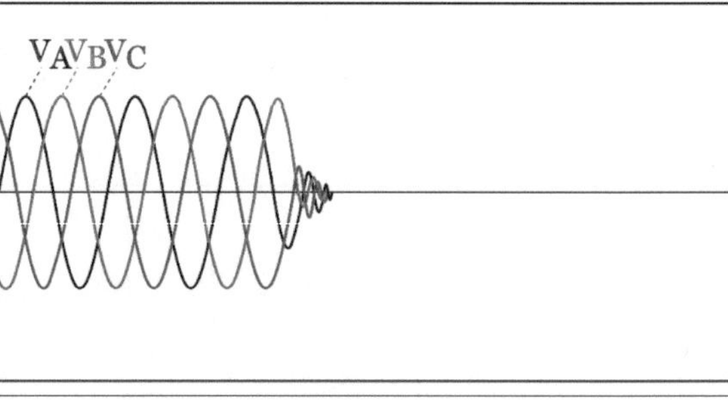

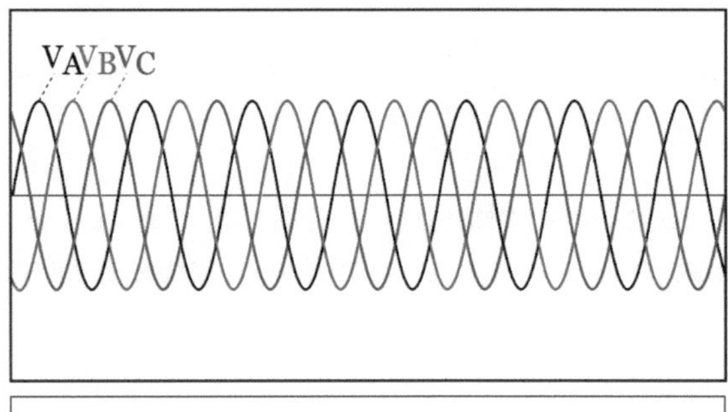

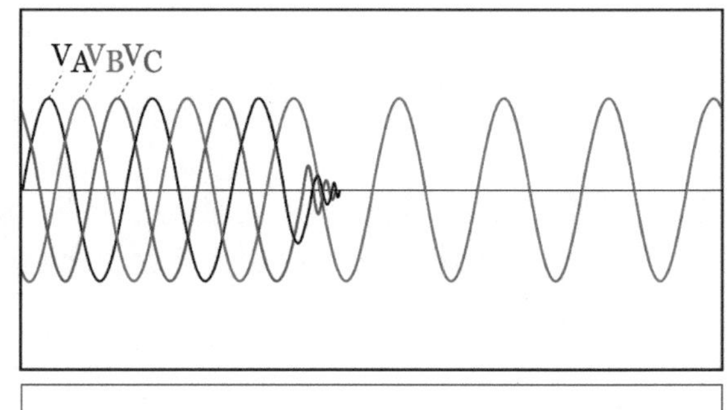

Normal Operating Conditions	Single-phase to Ground Fault
Phase to Phase Fault	Three Phase Fault

52. A delta-wye grounded transformer is rated for 115 kVA, 4.16k-480/277V, 6.7% impedance, and 89% efficiency. Select all statements below that apply. Assume balanced and positive ABC sequence conditions.

(A) The secondary line voltage will lead the primary line voltage by 30 degrees.

(B) The secondary line current will lead the primary line current by 30 degrees.

(C) There is no phase shift in the secondary line voltage or secondary line current compared to the primary line voltage and primary line current.

(D) The three-phase delta-wye transformer voltage ratio is smaller by a factor of the square root of three compared to the per-phase voltage ratio of each single-phase transformer inside of the three-phase transformer.

(E) Zero sequence current can flow through the transformer.

53. A 75 mile long single-phase transmission line is rated to deliver 1 MVA of power at 2.4 kV. Select all statements below that apply.

(A) The average line inductance is twice as large as the total line inductance.

(B) The average line inductive reactance is twice as large as the total line inductive reactance.

(C) The average line capacitance is twice as large as the total line capacitance.

(D) The average line capacitive reactance is twice as large as the total line capacitive reactance.

(E) Charging current is directly proportional to the total capacitive reactance.

54. In the three-phase single-line diagram shown below, the input power to generator #1 is 40 MW and the input power to generator #2 is 45 MW.

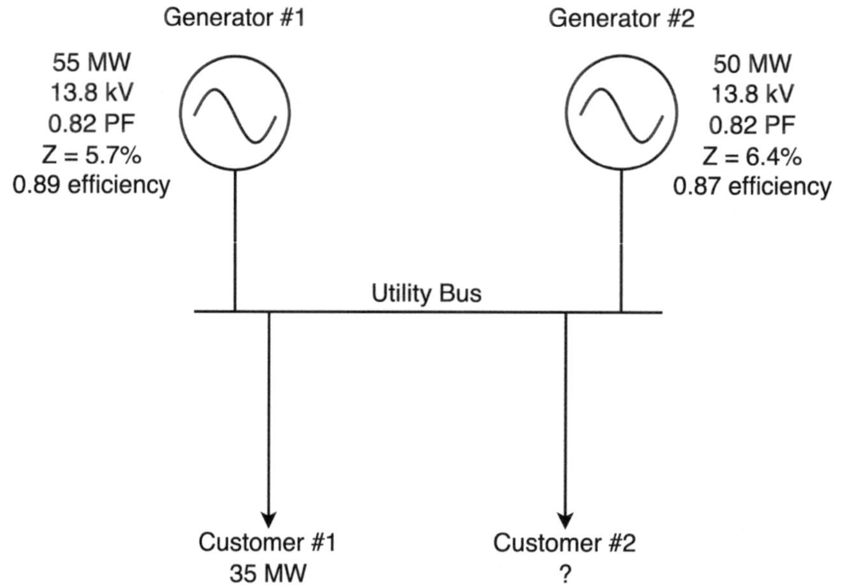

The amount of real power delivered to customer #2 to the nearest megawatt is _________ MW.

Fill in your response.

55. A utility synchronous generator with an inertia constant of 6 megajoules per megavolt-amperes (MJ/MVA) is operating at steady state synchronous speed and rated conditions with 1 pu of output power.

If the initial rotor electrical torque angle is 26 degrees, then the maximum amount of time to clear the fault before the generator becomes unstable is _________ seconds.

Fill in your response to the nearest two decimal places.

56. Select all statements below that apply to impedance and changes in frequency due to harmonics.

(A) Inductive reactance is directly proportional to frequency.

(B) Capacitive reactance is directly proportional to frequency.

(C) Resistance is directly proportional to frequency.

(D) Impedance is directly proportional to frequency.

(E) Reactance at the third harmonic will be three times as large compared to the fundamental frequency if it is inductive.

57. A 75 mile long three-phase transmission line is rated to deliver 220 kV at the receiving end. The total series impedance of the transmission line is 30 + j135 ohms with a shunt admittance of j750 microsiemens.

The surge impedance loading of the transmission line is _________ megawatts.

Fill in your response to the nearest unit shown.

58. Overcurrent protection is intended to interrupt which types of occurrences? Select all that apply.

(A) Short circuit

(B) Ground fault

(C) Overload

(D) Under frequency

(E) Voltage transients

59. Drag and drop the ANSI device number to the appropriate place on the single-line diagram. You may use each ANSI device number more than once.

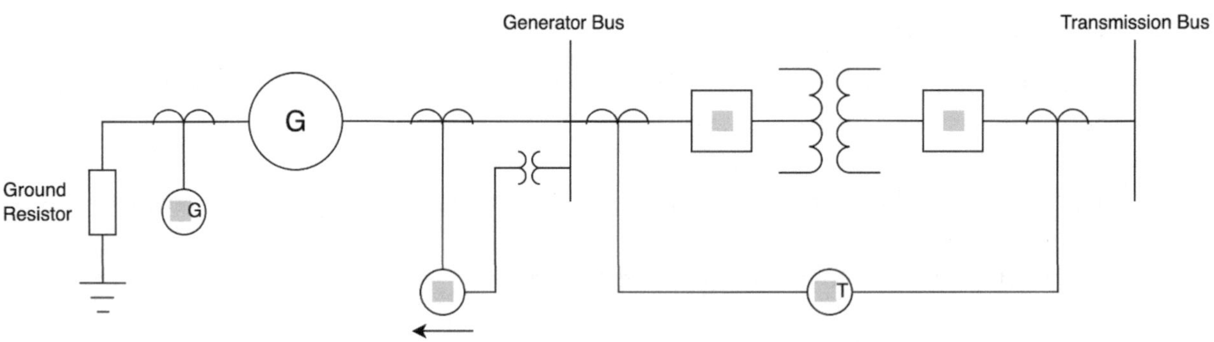

ANSI device numbers drag and drop:

87 52 51 67

60. Drag and drop each possible value below that best describes the overcurrent trip characteristics of the device shown below. Each value may be used more than once, and some values will not be used.

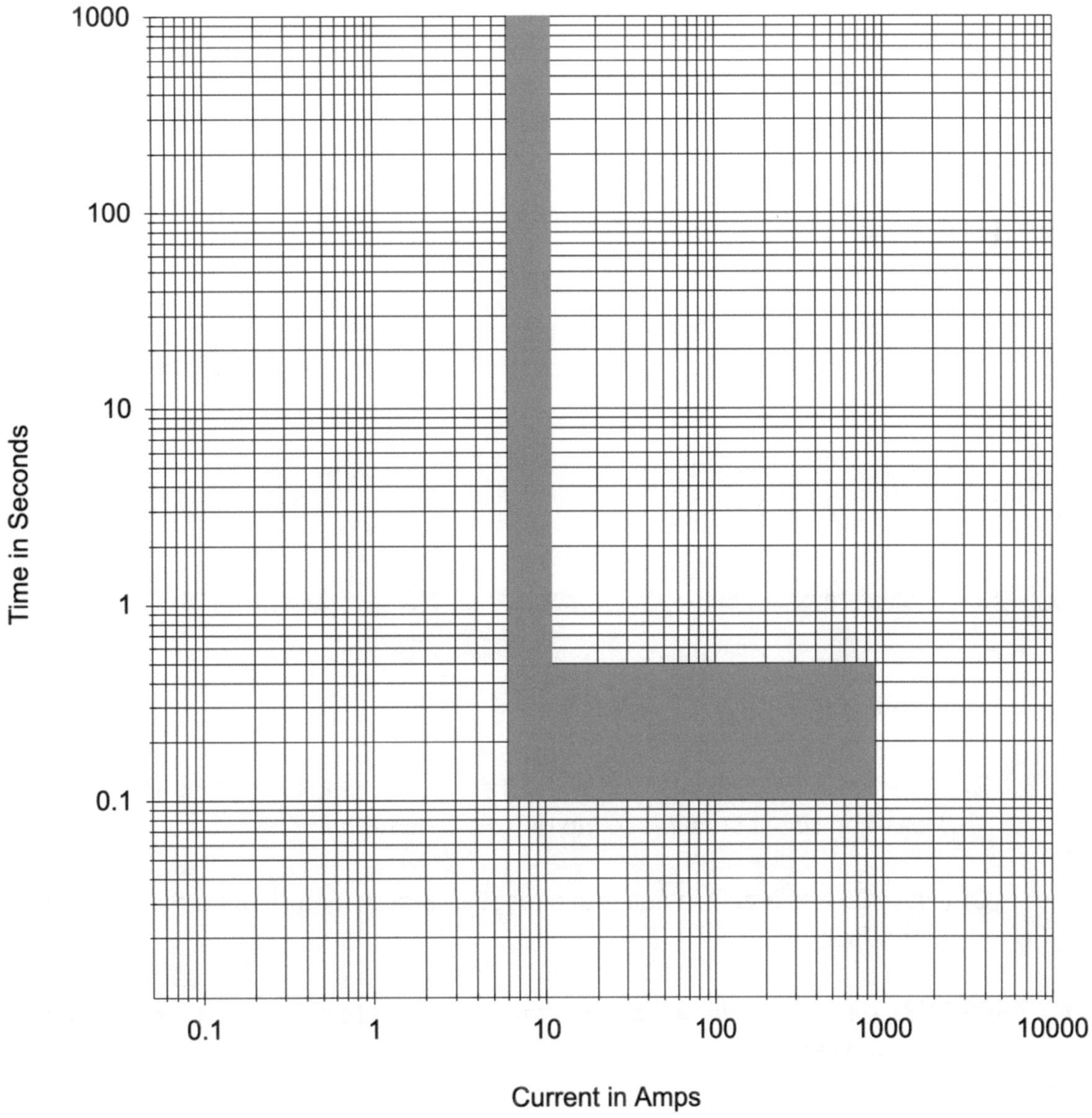

Pick up current: ☐ Amps　　Maximum fault current: ☐ Amps

Time delay: ☐ Cycles　　Maximum instantaneous clearing time: ☐ Cycles

2	6	12	20	30	100	200	300	600	900

61. Select all statements below that apply to core balance ground fault protection.

(A) One current transformer is used for all current carrying conductors

(B) Measures zero sequence current

(C) Measures negative sequence current

(D) Measures residual current

(E) Connects directly to the system grounded neutral

62. Select all statements below that apply to the ANSI #86 device.

(A) ANSI #86 is a protection relay.

(B) ANSI #86 is a multicontact relay used as an intermediary device between protection relays and circuit breakers to trip or isolate more than one circuit.

(C) ANSI #86 has a built-in automatic reset so that tripped circuit breakers may reclose after a fault.

(D) ANSI #86 must have one current transformer and one potential transformer as the minimum input.

(E) One application of the ANSI #86 device is detecting sensitive ground faults.

63. Drag and drop the name of each type of distance relay function to its appropriate R-X diagram shown below.

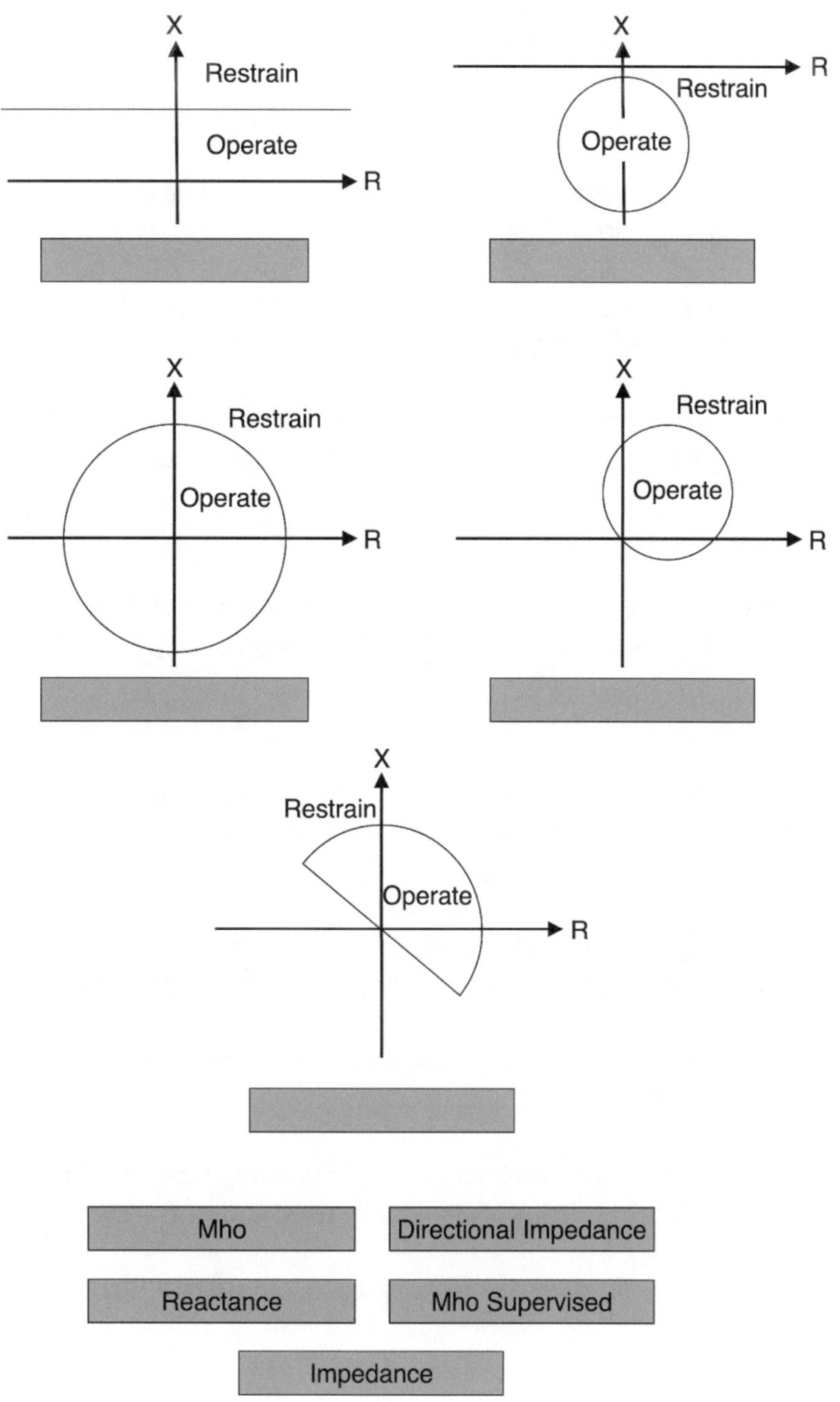

Mho	Directional Impedance
Reactance	Mho Supervised
Impedance	

64. Select each statement below that applies to low-voltage fuse selection and application based on class type.

(A) Class H fuses are intended for current limiting applications and are available in 250 volt or 600 volt ratings up to 600 amps.

(B) Class R, Class J, and Class L fuses are intended for current limiting applications.

(C) Class H and Class K fuses are interchangeable in physical dimensions, but Class H has greater interrupting ratings.

(D) Dual-element time-delay fuses may be used to protect against short circuit overcurrents while permitting normal temporary overloads.

(E) Current limiting fuses may be used to protect downstream circuit breakers that have an interrupting rating less than the available short circuit current.

65. Select each statement below that applies to molded-case circuit breakers (MCCBs), insulated-case circuit breakers (ICCBs), and low-voltage power circuit breakers (LVPCBs).

(A) Molded-case circuit breakers are fast interrupting circuit breakers with all components built into a single insulating molded case that are mostly fixed-mounted and typically operated with a mechanical toggle.

(B) Insulated-case circuit breakers are larger in frame size compared to molded-case circuit breakers and feature a stored energy operating mechanism for operation.

(C) Insulated-case circuit breakers are most commonly used for residential and commercial lighting and receptacle circuits.

(D) Low-voltage power circuit breakers are larger in frame size compared to insulated-case circuit breakers and primarily used in drawout switchgear.

(E) Low-voltage power circuit breakers are rated to interrupt 100% of their continuous current rating.

66. An ANSI #50P relay is fed from a 600:5 multi-ratio current transformer (MRCT) using the secondary X3 and X4 MRCT taps. The relay is programmed for a 10 amp tap setting which results in 50 volts across the MRCT core during operation. Which of the following fault currents will cause the relay to operate, assuming that the CT does not saturate? Select all that apply.

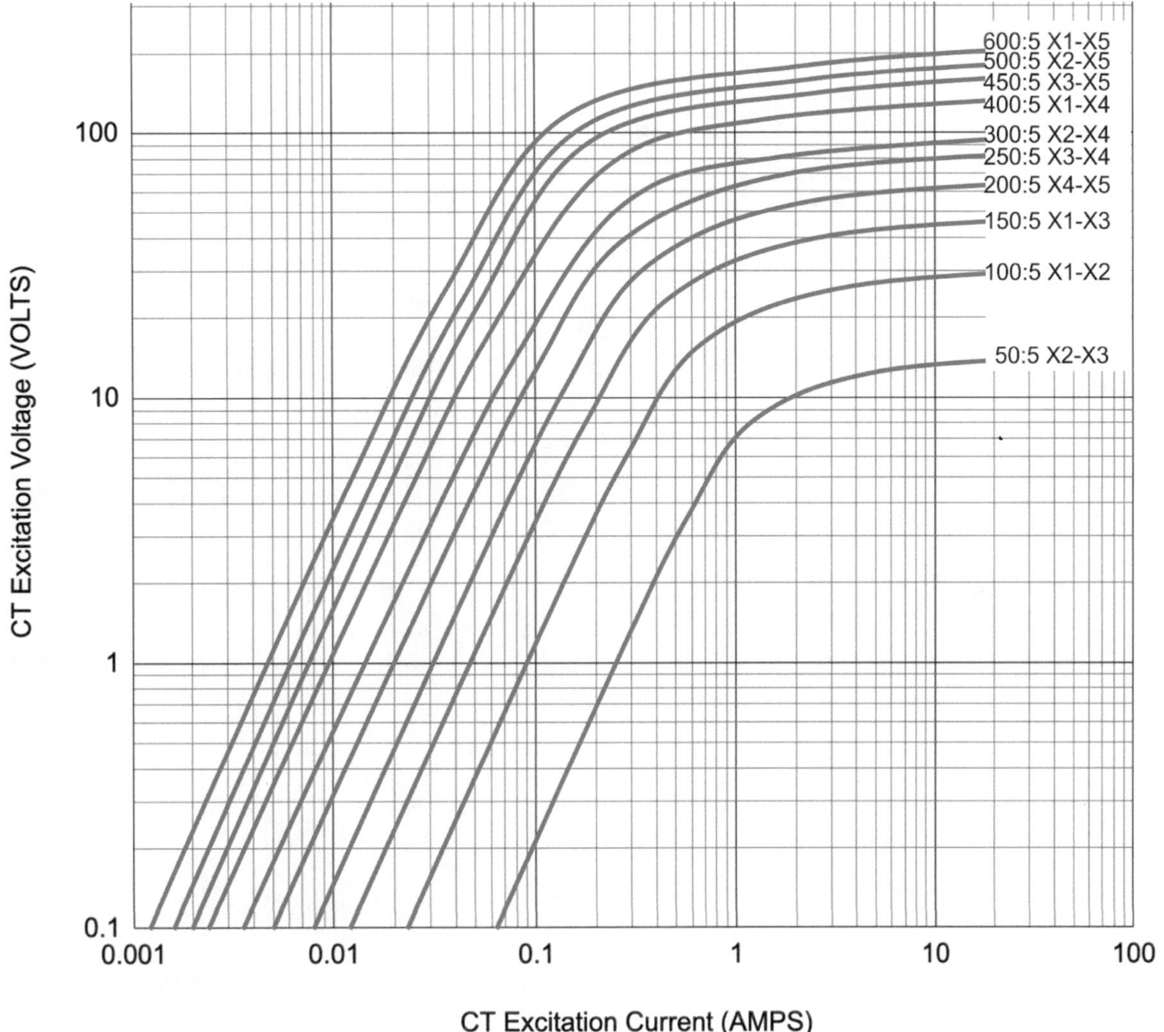

(A) 600 amps
(B) 550 amps
(C) 500 amps
(D) 450 amps
(E) 400 amps

67. Drag each device label to its appropriate time current characteristic (TCC) curve shown below.

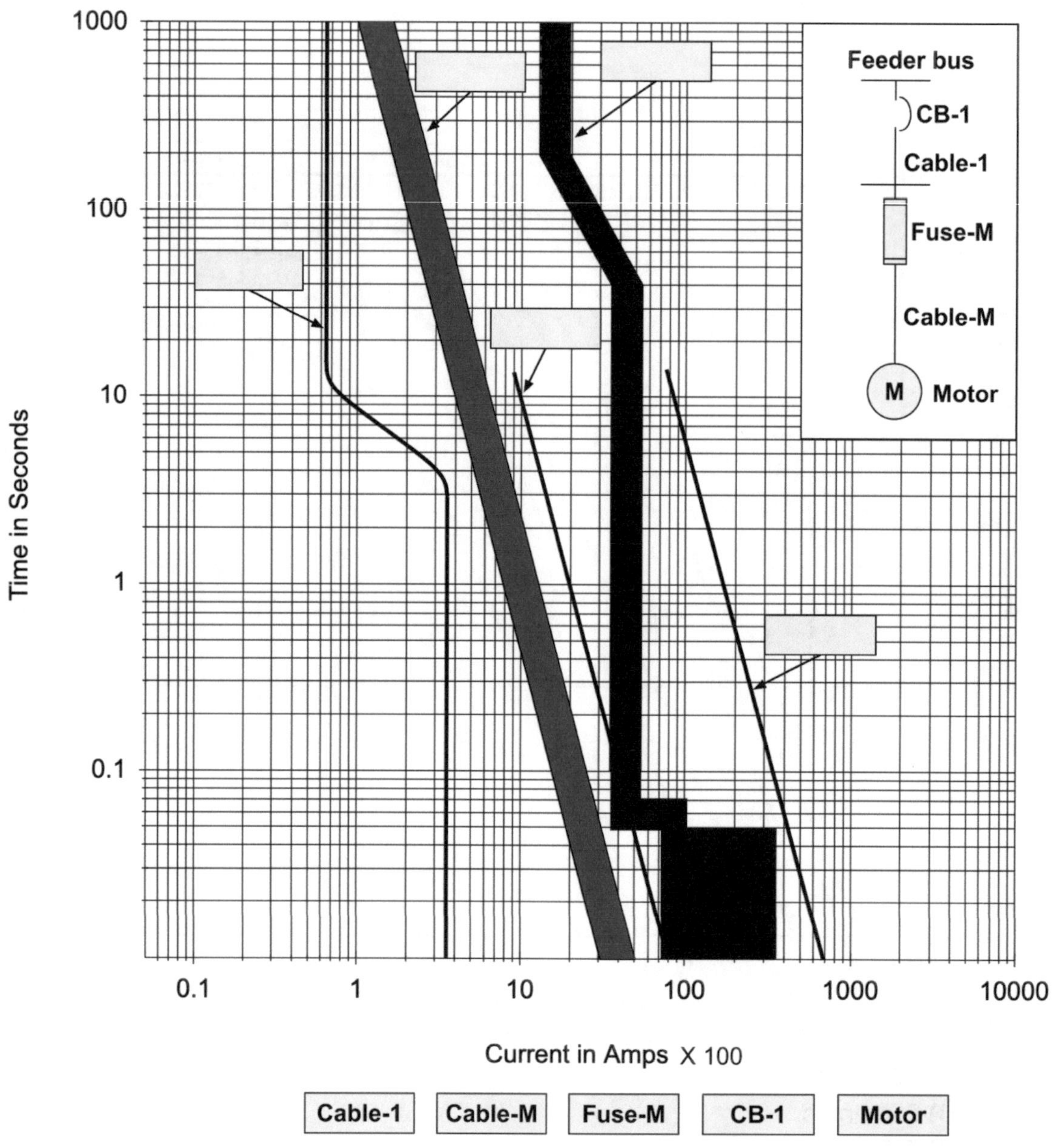

68. Use the low-voltage power circuit breaker (LVPCB) time current characteristic (TCC) curve shown below to approximate the following values. Fill in your responses.

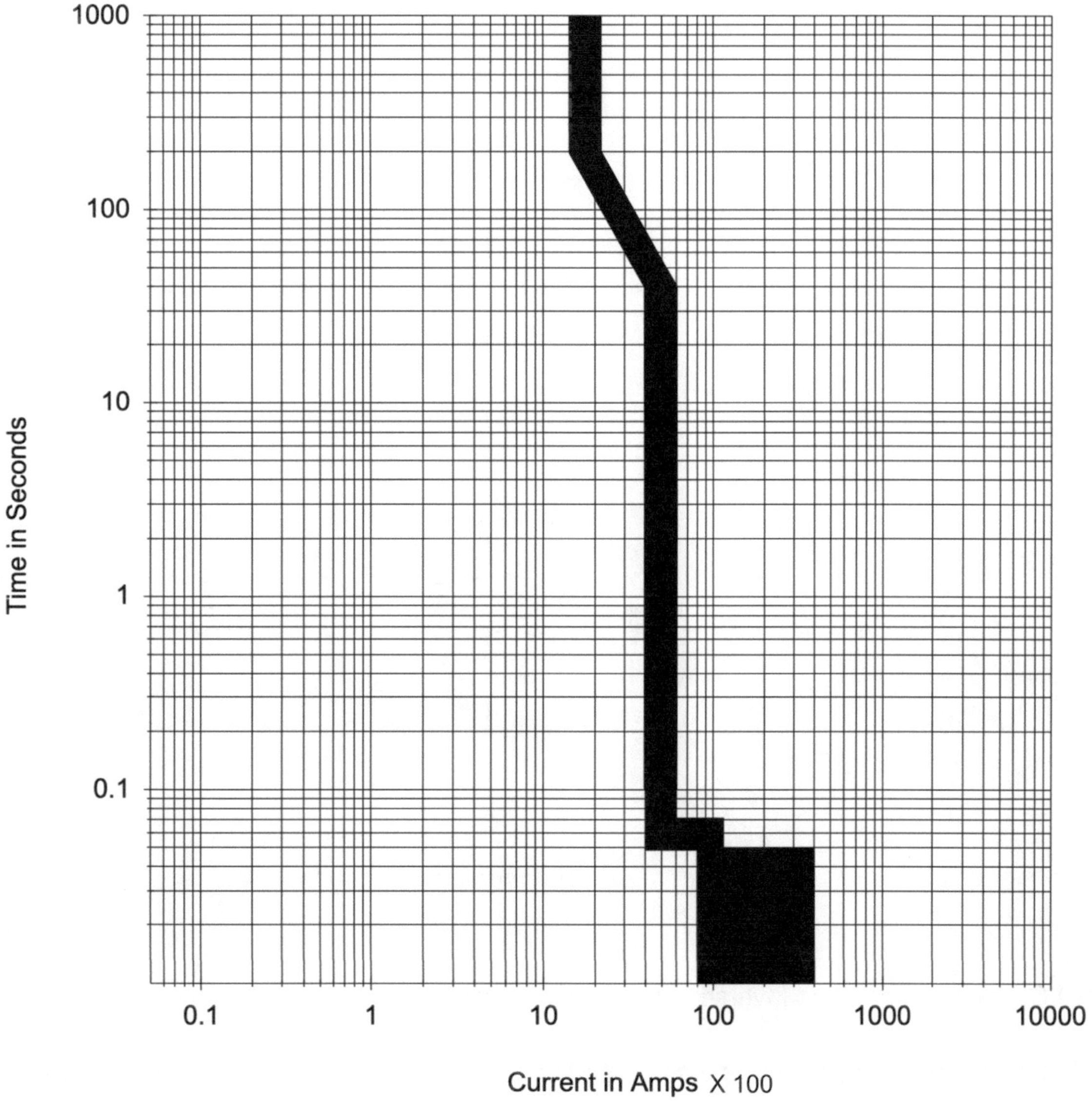

Time in Seconds

1000
100
10
1
0.1

0.1 1 10 100 1000 10000

Current in Amps X 100

The long-time pickup current is _________ amps.

The short-time pickup current is _________ amps.

The instantaneous pickup current is _________ amps.

The maximum interrupting current is _________ amps.

69. Select all statements that apply to the two time current characteristic (TCC) curves shown below.

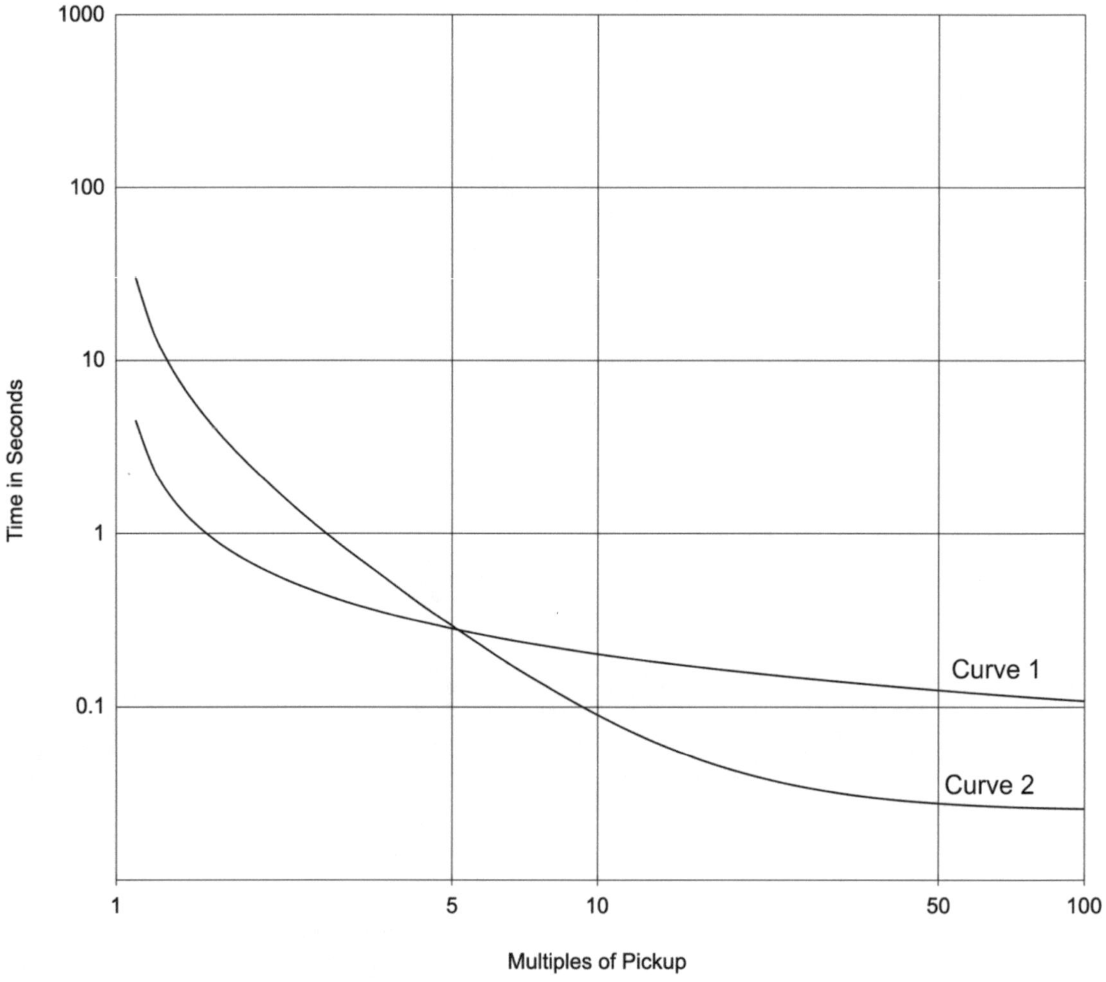

(A) Curve 1 is more time inverse compared to curve 2.

(B) Curve 2 is more time inverse compared to curve 1.

(C) Both curves are examples of ANSI #51 relay functions.

(D) Both curves are examples of ANSI #50 relay functions.

(E) Curve 1 and curve 2 are examples of different time dial settings.

70. Correctly label each type of hazard location shown below according to the appropriate *2017 National Electrical Code®* class and division by dragging and dropping each label. Each label may only be used once.

(A) A textile mill that produces fabric containing cotton fibers that are easily ignitible.

(B) A grain milling facility that produces combustible dust with the potential of explosion only if the grain milling machinery fails.

(C) A textile storage facility that stores and transports cotton fibers that are easily ignitible.

(D) A dry cleaning plant where combustible liquid-produced vapors are present in quantities sufficient to produce a combustion only when the cleaning machinery is undergoing routine maintenance.

(E) A grain processing facility that handles combustible dust that settles over time on top of electrical equipment which, if not routinely cleaned, causes 480 volt power and lighting transformers to overheat.

(F) A chemical storage facility that receives, stores, and transports dry cleaning liquids in enclosed airtight containers. The liquid chemicals produce combustible vapors in sufficient quantities that would result in combustion if the containers were to fail.

Class I Division 1	Class II Division 1	Class III Division 1
Class I Division 2	Class II Division 2	Class III Division 2

71. Select each statement below that applies to Class I Division 1 hazardous locations according to the *2017 National Electrical Code®*. Select all that apply.

(A) Wall-mounted PVC conduit is permitted for Class I Division 1 hazardous locations.

(B) Wiring methods that are allowable for Class I Division 1 hazardous locations are also permitted for use in Class I Division 2 hazardous locations.

(C) Hermetically sealed electrical enclosures do not require conduit seals in Class I Division 1 hazardous locations.

(D) General-purpose electrical enclosures may be used for surge protective devices in Class I Division 1 hazardous locations as long as they are non arcing.

(E) Only 90°C temperature rated conductors are permitted for Class I Division 1 hazardous locations.

72. Which of the following wire sizes are permitted according to the *2017 National Electrical Code®* for a 480 volt, three-phase load, 125 amp circuit if three (3) THWN copper current-carrying conductors are run in PVC conduit with an ambient temperature of 110 degrees Fahrenheit? Select all that apply.

(A) 2 AWG

(B) 1 AWG

(C) 1/0 AWG

(D) 2/0 AWG

(E) 3/0 AWG

73. Which of the following wire sizes are permitted according to the *2017 National Electrical Code®* for two TW aluminum current-carrying conductors in rigid metal conduit (RMC) for a 50 amp, 120 volt circuit if there are already six conductors in the conduit? Select all that apply.

(A) 8 AWG

(B) 6 AWG

(C) 4 AWG

(D) 3 AWG

(E) 2 AWG

74. Determine each statement below that is permissible according to the *2017 National Electrical Code®*. Select all that apply.

(A) The maximum standard overcurrent device rating for a branch circuit supplying a 208 volt 175 amp continuous load fed from the minimum required wire size of not more than three 75ºC rated copper current-carrying conductors in raceway is 250 amps.

(B) 225 amps is the maximum standard overcurrent device rating that may be used to protect a conductor with an allowable ampacity of 200 amps.

(C) A three-phase 480 volt panel is fed from two THWN copper 700 kcmil conductors per phase. Each set of three current-carrying conductors is run in separate conduit. 1,000 amps is the maximum standard overcurrent device rating that may be used to protect this circuit.

(D) A three-phase 480 volt panel is fed from two THWN copper 700 kcmil conductors per phase. Each set of three current-carrying conductors is run in separate conduit. An inverse time circuit breaker rated for 900 amps supplied by a manufacturer may be used to protect this circuit.

(E) The smallest copper-clad aluminum 75ºC rated wire size that can be protected by an 80 ampere rated overcurrent device is 3 AWG for a circuit rated up to 2,000 volts with not more than three current-carrying conductors in conduit.

75. Below is the nameplate on the enclosure of an outdoor residential HVAC condenser with a hermetically sealed motor-compressor and a condenser fan motor.

SUITABLE FOR OUTDOOR USE		
POWER SUPPLY:		
1 PH	60 HZ	208-230 VAC
COMPRESSOR:		
13.50 RLA	72.5 LRA	60 HZ
FAN MOTOR:		
1/4 HP	1.40 FLA	60 HZ
18.3 MINIMUM CIRCUIT AMPS		
30 A MAX FUSE/CKT BKR		
5 KA RMS SYMMETRICAL MAX SHORT CIRCUIT		

Which of the following is applicable according to the *2017 National Electrical Code®*? Select all that apply.

(A) The minimum 60ºC copper wire size for this multimotor equipment is 14 AWG.

(B) A 25 amp circuit breaker is permitted for overcurrent protection.

(C) The minimum full-load current equivalent disconnect rating permitted, rounded to the nearest ampere, is 17.

(D) The disconnect for this condenser must be within 30 feet of the equipment.

(E) A motor controller with a 10 kA short-circuit current rating (SCCR) is permitted.

76. Which of the following choices below are true and permitted for cable tray installations according to the *2017 National Electrical Code®*? Select all that apply.

(A) Cable tray is considered as a raceway when determining conductor allowable ampacity ratings for single conductors.

(B) Cable tray is only permitted to be installed in industrial locations.

(C) Parallel circuit conductors must be installed in separate cable trays and bundled together only at the power supply and load terminals.

(D) Where all cables are multiconductor size 4/0 AWG and larger, the sum of the diameters of all cables may not exceed 50% of ladder-type cable tray width.

(E) Metal cable trays are permitted to be used as equipment grounding conductors in locations where continuous maintenance and supervision ensure that only qualified persons service the cable tray systems.

77. Which of the following choices below are true and permitted for receptacles according to the *2017 National Electrical Code®?* Select all that apply.

(A) The rating of a branch circuit is determined by the allowable ampacity rating of the branch-circuit conductor.

(B) A single receptacle installed on an individual branch circuit must have an equal or greater ampere rating than the overcurrent protection device of the circuit.

(C) A branch circuit protected by a 20 ampere rated overcurrent protection device may supply multiple receptacles rated for both 15 amperes and 20 amperes.

(D) 15 amperes is the maximum rating for a cord-and-plug equipment powered by a receptacle on a 20 ampere rated branch circuit.

(E) Branch circuits in a multiple occupancy dwelling unit may not supply loads in an adjacent dwelling unit.

78. Which of the following choices below are true and permitted for feeder conductors according to the *2017 National Electrical Code®?* Select all that apply.

(A) If adjustment or correction factors are not required for a feeder up to 600 volts, then the allowable ampacity must be equal to or greater than the sum of 125% of the continuous loads and 100% of the noncontinuous loads.

(B) If adjustment or correction factors are required for a feeder up to 600 volts, then the allowable ampacity must be equal to or greater than the maximum load.

(C) If the neutral conductor of a feeder up to 600 volts is not connected to an overcurrent device, it must have an allowable ampacity of at least the sum of 100% of the noncontinuous load and 100% of the continuous load.

(D) The minimum rating for feeder overcurrent protection is the sum of the continuous loads and noncontinuous loads.

(E) Up to two sets of 3-wire feeders may share a common neutral conductor.

79. The maximum number of 20 ampere rated or less single-yoke duplex receptacles permitted on a 20 ampere rated 120 volt general-purpose branch circuit in a commercial occupancy according to the *2017 National Electrical Code®* is _________.

Fill in your response.

80. Drag and drop the labels shown below to their correct locations according to the *2017 National Electrical Code®.*

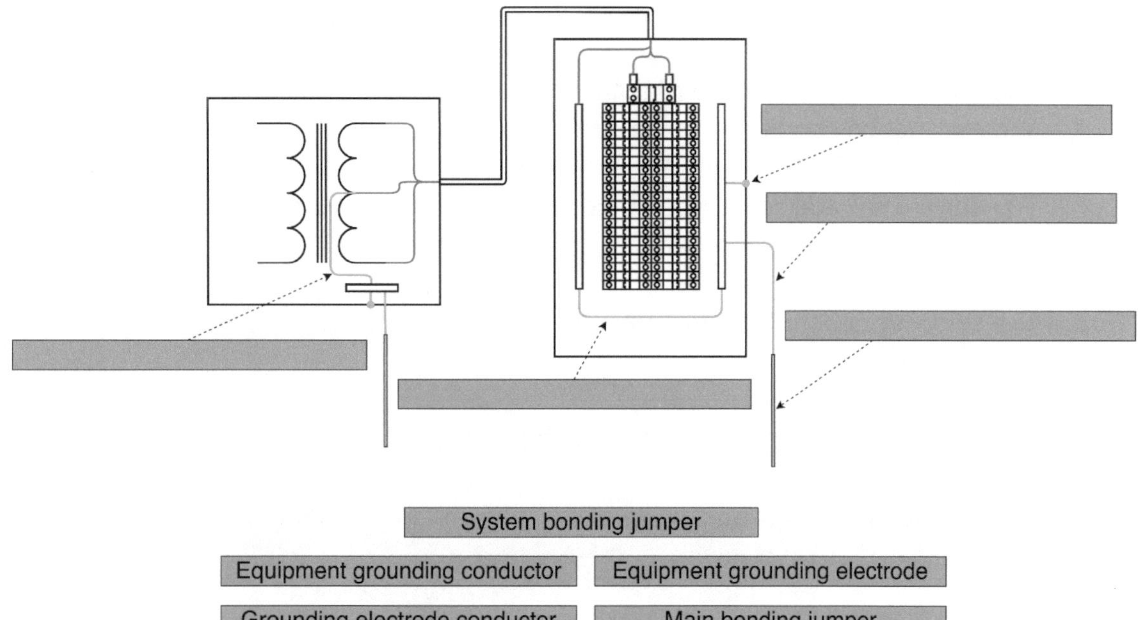

System bonding jumper

Equipment grounding conductor | Equipment grounding electrode

Grounding electrode conductor | Main bonding jumper

Problem #1 Solution

The following rules must be met in order for the two-wattmeter method to accurately measure the power of a three-phase, three-wire, unbalanced system:

1. The negative voltage polarity of both wattmeters must be connected to the same phase.

2. The positive voltage polarity of both wattmeters must be connected to different phases and may not be connected to the same phase as the negative voltage polarity.

3. The current coil that measures amperage must be connected to the same phase as the positive voltage polarity for each wattmeter.

4. The current coil of both wattmeters must measure amperage in the direction from the source to the load.

In this problem, all five configurations meet all four rules.

The positive voltage polarity is designated by a "+" sign, and the direction of current flow being measured is indicated by an arrow.

The source and load were not identified, but it can be assumed that the source is on the left and the load is on the right since all five configurations show the current flow measurement from left to right.

The answer is: **A, B, C, D, and E.**

(Multiple correct AIT question) - Ch. 1.1 Instrument Transformers

Problem #2 Solution

Since the relay and current transformer (CT) are on the primary side of the transformer, we will need to calculate 100% of the transformer's primary full load amps first before stepping this value down. Because this is a fill in the blank AIT question, we will carry all values in our calculator and only round at the very end of the solution.

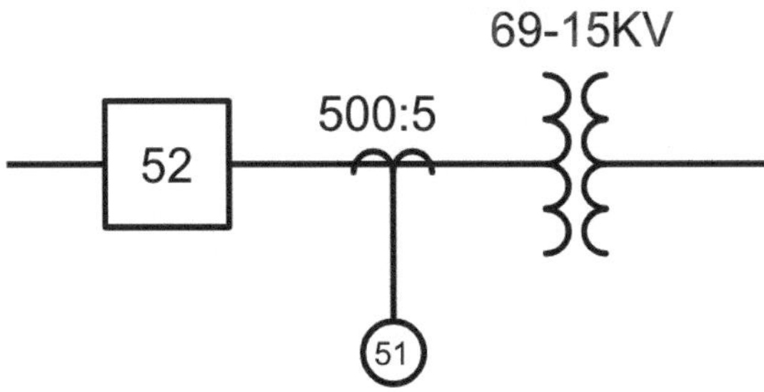

We can solve for 100% of the three-phase, 55 MVA (given in the problem) transformer's primary full load amps using the three-phase apparent power formula and solving for line current:

$$|I_L| = \frac{|S_{3\emptyset}|}{\sqrt{3} \cdot |V_L|}$$

DEG

$$\frac{55E6}{\sqrt{3}*69E3}$$

460.2067363

100% of the transformer's primary full load amps is 460 A.

Next, we need to step down this value using the CT ratio to determine the amount of current fed to the ANSI #51 relay.

CTs always **step down** current, so be sure to multiply the line current by the ratio of the smaller number of CT turns to the larger number of the CT turns.

Using the answer from the previous step in the calculator and a CT ratio of 500:5, we can calculate the secondary CT current leaving the CT and entering the relay:

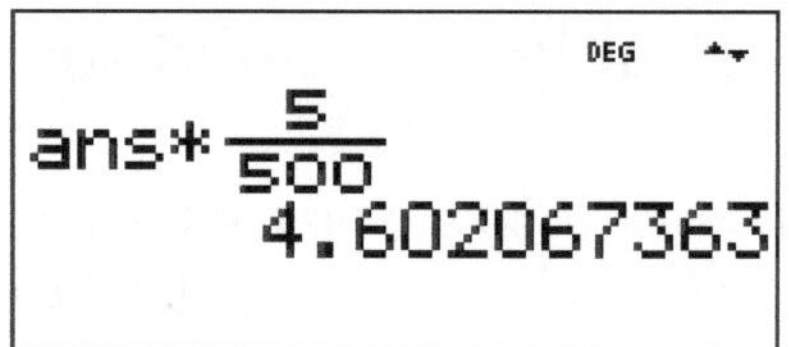

Rounding to the nearest ampere gives 5 amps.

The answer is: **5.**

(Fill in the blank AIT question) - Ch. 1.1 Instrument Transformers

Problem #3 Solution

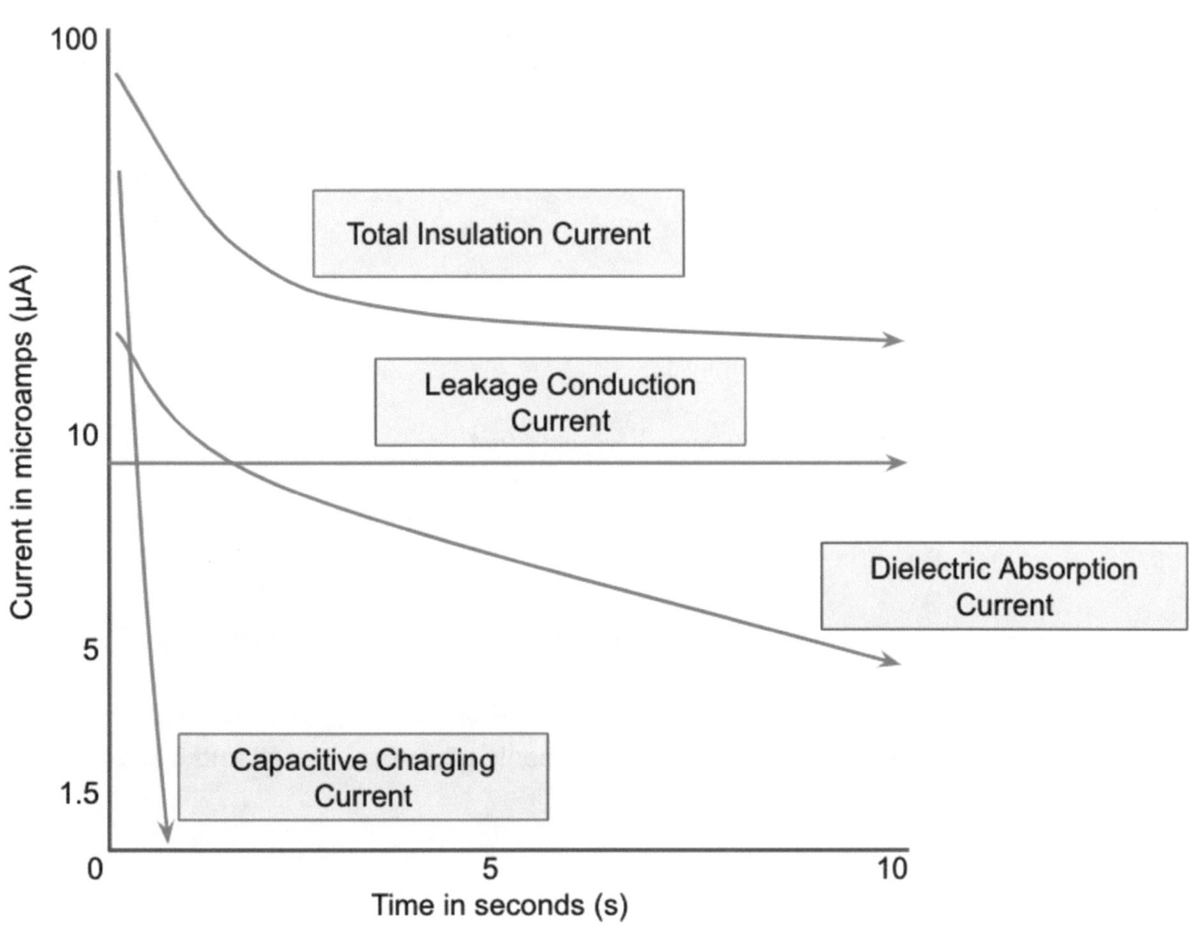

The exact values of each of the insulation currents with respect to time and their resulting graph on the time scale will vary with the type of insulation being tested and the condition of the insulation. In general, however, each of the currents will behave similarly with respect to one another as the curves in the above graph show. A similar graph appears in the *NCEES® Reference Handbook*[1].

The answer is: **see filled in labels in graph above.**

(Drag and drop AIT question) - Ch. 1.2 Insulation Testing

[1] *NCEES® Reference Handbook (Version 1.1.2) - 2.1.1.2 Characteristics of the Measured Direct Current p. 17*

Problem #4 Solution

Insulation resistance varies inversely with the exponential of the conductor temperature at the time of measurement. Because of this, most criteria for recommended minimum insulation resistance (pass vs fail) is based on insulation resistance measured at a standard of 40 degrees Celsius.

We can temperature correct the insulation resistance measured at any temperature to 40 degrees Celsius using the following formula[2]:

$$R_I(40^oC) = k(T^oC) \cdot R_I(T^oC)$$

where:

$R_I(40^oC)$ = The insulation resistance corrected to 40 degrees Celsius.
$k(T^oC)$ = The temperature correction coefficient.
$R_I(T^oC)$ = The insulation resistance measured at a temperature other than 40 degrees Celsius.

To use the formula above, we'll first need to calculate the temperature correction coefficient (k).

For **thermoset** (and thermosetting) insulation, the temperature correction coefficient (k) can be calculated using the following formula[3] for measurements taken between **40 and 85 degrees Celsius**:

$$k(T^{\underline{o}}C) = exp\left[-\ 4230\left(\frac{1}{T+273} - \frac{1}{313}\right)\right]$$

where:

exp = The exponential constant e. For example: $exp[x] = e^x$.
T = The temperature in Celsius that the insulation was measured at.

Let's calculate the temperature correction coefficient (k) for thermoset insulation using **T = 65 degrees Celsius**, the temperature that the insulation resistance was measured at according to the problem:

[2] *NCEES® Reference Handbook (Version 1.1.2) - 2.1.1.3 Effect of Temperature p. 18*
[3] *NCEES® Reference Handbook (Version 1.1.2) - 2.1.1.3 Effect of Temperature p. 18*

```
                DEG
  (-4230( 1/(65+273) - 1/313 ))
e
        2.717151523
```

The temperature correction coefficient (k) is approximately 2.717.

Next, let's use the temperature correction coefficient (k) to correct the 125 megaohm insulation resistance measured at 65 degrees Celsius to the resistance at 40 degrees Celsius:

$$R_I(40^oC) = k(T^oC) \cdot R_I(T^oC)$$

```
                DEG
ans*125
        339.6439404
```

The insulation resistance corrected to 40 degrees Celsius, to the nearest megaohm [MΩ], is 340.

The answer is: **340.**

(Fill in the blank AIT question) - Ch. 1.2 Insulation Testing

Problem #5 Solution

Since this is a fill in the blank question, we will be carrying our calculations in the calculator until the final answer when we will round.

Four grounding rods are being used with unequal spacing, so we safely assume the **Schlumberger Method**[4] is being used to measure the resistance to ground:

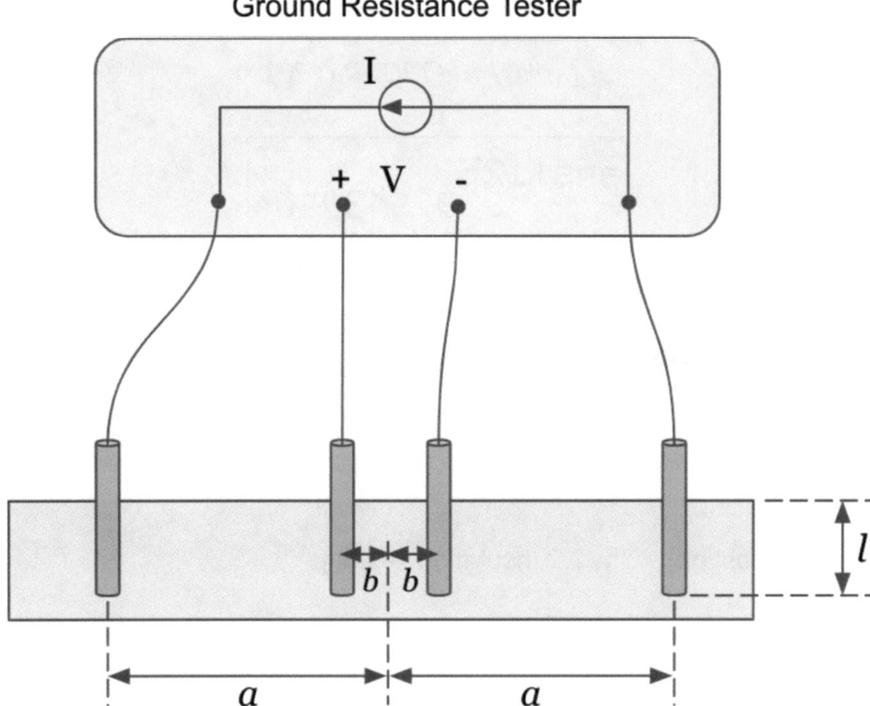

Since the outer two rods are 300 feet apart, we can calculate the distance a:

$$a = \frac{300\ ft}{2} = 150\ ft$$

Since the inner two rods are 50 feet apart, we can calculate the distance b:

$$b = \frac{50\ ft}{2} = 25\ ft$$

Since a is much larger than two times the distance of b (a >> 2b), we'll use the simplified Schlumberger Method formula:

[4] *NCEES® Reference Handbook (Version 1.1.2) - 2.1.2 Ground Resistance Testing p. 19*

$$\rho = \frac{\pi a^2 R}{2b}$$

Rearranged to solve for resistance (R), the formula becomes:

$$R = \frac{\rho 2b}{\pi a^2}$$

> ***Careful!*** *The given soil resistivity (ρ) is in ohm centimeters [Ω•cm] and the two rod distances (a and b) are in feet. In order to get these to cancel, we'll need to convert these values to either all centimeters or all feet.*

Since there is one centimeter value but two feet values, it will be quicker to convert the soil resistivity (ρ) to ohm feet [Ω•ft]:

ρ = (25,000Ω•cm)(3.281×10^{-2} ft/cm) = 820.25 Ω•ft

```
DEG
25E3*3.281E-2
          820.25
```

Now let's calculate the resistance to ground (R):

```
DEG
ans*2*25
--------
 π*150²
     0.580208187
```

Rounded to one decimal place like the question asks for means that the resistance to ground in ohms [Ω] is approximately 0.6.

The answer is: **0.6.**

(Fill in the blank AIT question) - Ch. 1.3 Ground Resistance Testing

Problem #6 Solution

Let's compare each of the possible answer choices to determine which of the following are advantages unique to the **Variation of Depth Method (Driven Rod Method)**[5]:

(A) The ability to take a ground resistance measurement at the testing location.

All ground resistance testing methods are used to take a ground resistance measurement at the testing location. This is **not** unique to the driven rod method.

A is false.

(B) Only three rods are required for the test compared to four.

The driven rod method, also known as the variation of depth method, only uses three ground rods:

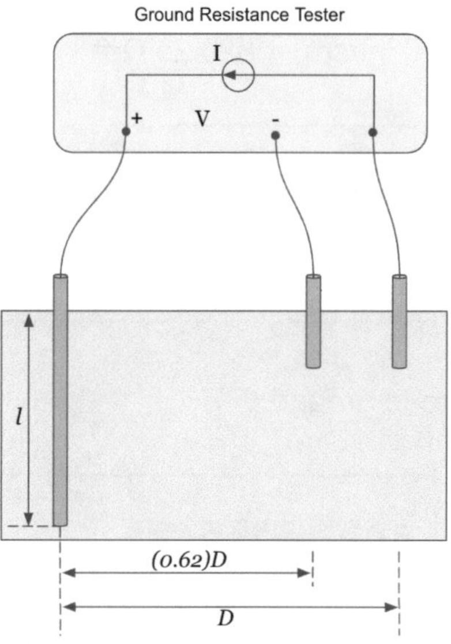

In comparison, most other popular ground testing methods use four rods.

B is true.

(C) The ability to determine how far permanent ground rods will be able to be driven in the testing location.

Ground testing rods (or "pins") are typically inserted into the same shallow depth distance (*l*)

[5] *NCEES® Reference Handbook (Version 1.1.2) - 2.1.2 Ground Resistance Testing p. 20*

for most popular ground testing methods except for the driven rod method which requires the main rod to be driven much deeper into the soil.

Because of this, the driven rod method has the advantage of immediately discovering the soil conditions at the ground testing site in terms of soil density and difficulty of driving permanent ground rods when a suitable low resistance to ground location is found.

For example, with four pin methods like the Schlumberger or Wenner methods which do not drive the testing rods very deep into the soil, it is possible to find a suitable low resistance to ground location to install permanent grounding rods without realizing there could be very dense soil conditions such as coral or rock under the testing location. These conditions can lead to extra difficulty during the permanent ground rod installation.

C is true.

(D) Greater accuracy in ground measurement compared to non-driven test methods.

Great force is required to drive the main testing rod deep into the soil during the driven rod method, usually with a hammer, rotary hammer equipment, or other impact methods. This induces numerous vibrations into the main test rod while it is being driven, possibly resulting in loose contact with the surrounding soil.

Because of this, the actual resistance to ground measurement recorded during the driven rod test is typically less accurate than other multiple pin non-driven tests and results in an artificially higher ground resistance measurement than what is actually present in the soil at the test site. This is the main disadvantage of the driven rod method.

D is false.

(E) All three testing rods may be equal in length compared to other methods.

With the driven rod method, the driven rod is much longer compared to the other two non-driven rods since it is driven deep into the soil at the test site. Non-driven ground tests that use more than one test rod (multiple pins) are typically all equal in length.

E is false.

The answer is: B and C.

(Multiple correct AIT question) - Ch. 1.3 Ground Resistance Testing

Problem #7 Solution

The **structural collection area (A_D)** of a building is used to determine the building's annual lightning strike frequency according to the *NFPA® 780 - Standard for the Installation of Lightning Protection Systems.*

This value can be calculated for a rectangular building using just the length (L), width (W), and height (H) of the building[6]:

$$A_D = LW + 6H(L+W) + \pi 9H^2$$

The building properties given in the problem are:

$L = 75\,ft$

$W = 50\,ft$

$H = 30\,ft$

We can calculate area (A_D) using the calculator:

```
DEG ▲                DEG ▲
75*50+6*30(75+50)+π9*30²
```

```
DEG ▲▼
75*50+6*30(75+5▸
      51696.90049
```

The collection area rounded to the nearest square foot is 51,697.

The answer is: **51,697.**

(Fill in the blank AIT question) - Ch. 2.1 Lightning Protection

[6] *NCEES® Reference Handbook (Version 1.1.2) - 2.2.1 Lightning Protection p. 21*

Problem #8 Solution

For **series reliability block diagram components**, in order for the system to be operable, ALL assets must be operable. Series block diagram assets are not redundant. This is represented by the formula[7]:

$$R(P_1, P_2, \ldots P_n) = \prod_{i=1}^{n} P_i$$

For **parallel reliability block diagram components**, in order for the system to be operable, at least one asset must be operable. Parallel assets are redundant. This is represented by the formula[8]:

$$R(P_1, P_2, \ldots P_n) = 1 - \prod_{i=1}^{n} (1 - P_i)$$

The block diagram in the problem has two series block diagram assets (1 and 3), and three parallel block diagram assets (2A, 2B, and 2C):

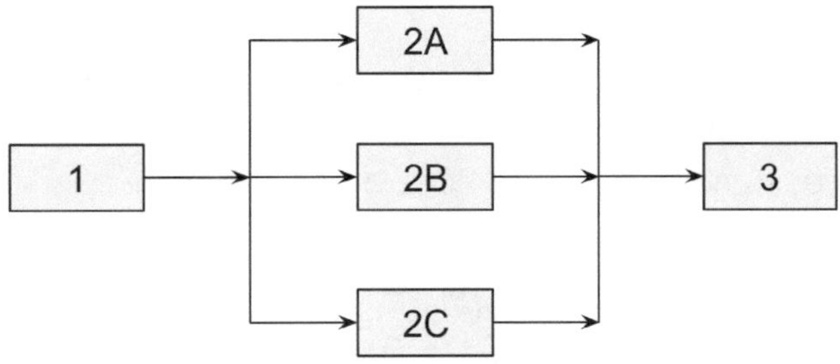

Let's evaluate each of the five possible answer choices:

(A) In order for the system to be operable, all five assets in the reliability block diagram must also be operable.

While both of the series block diagram assets (1 and 3) **must** be operable for the system to be operable, **only one** of the three parallel block diagram assets (2A, 2B, and 2C) must be operable in order for the system to be operable.

This means that the **minimum** number of assets that must be operable in order for the system to be operable is just three out of the five total: Assets 1, 3, and either 2A, 2B, or 2C.

---

[7] *NCEES® Reference Handbook (Version 1.1.2) - 2.2.2 Reliability p. 22*
[8] *NCEES® Reference Handbook (Version 1.1.2) - 2.2.2 Reliability p. 22*

Not all five assets shown are required to be operable in order for the system to be operable.

A is false.

(B) The system will be operable if assets 1, 2A, and 3 are operable.

Let's trace it out and assume that 2B and 2C are **not** operable:

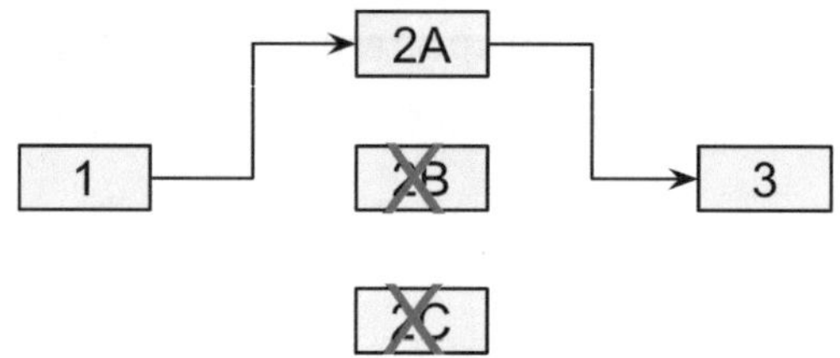

The system is still operable because it can get from the first input (asset 1) to the final output (asset 3).

B is true.

(C) The system will be operable if assets 1, 2B, and 3 are operable.

Let's trace it out and assume that 2A and 2C are **not** operable:

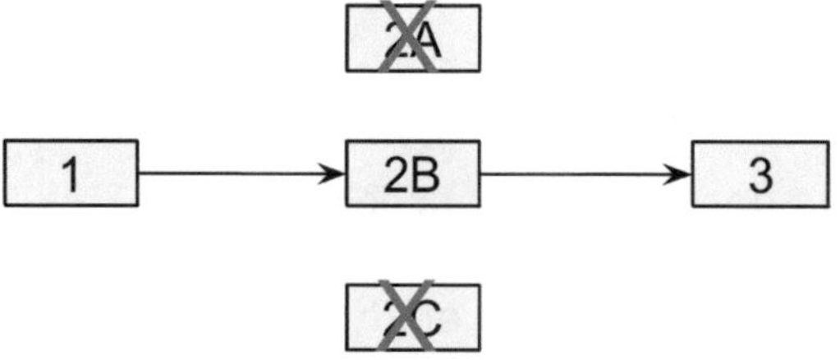

The system is still operable because it can get from the first input (asset 1) to the final output (asset 3).

C is true.

(D) The system will be operable if assets 1, 2B, 2C, and 3 are operable.

Let's trace it out and assume that 2A is **not** operable:

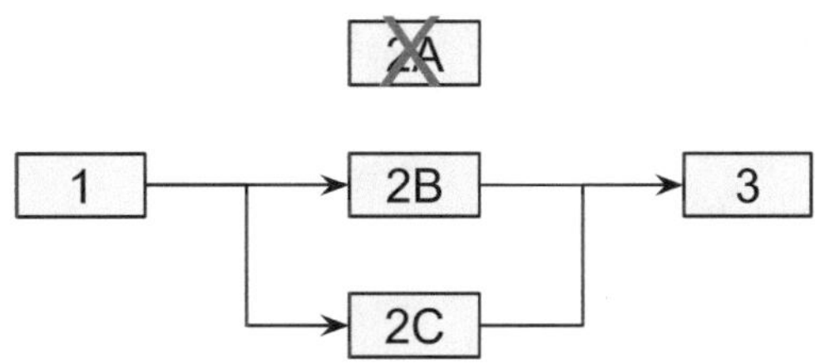

The system is still operable because it can get from the first input (asset 1) to the final output (asset 3).

D is true.

(E) The system will be operable if assets 1 and 3 are operable.

Let's trace it out and assume that 2A, 2B, and 2C are **not** operable:

The system is **not** operable because it **cannot** get from the first input (asset 1) to the final output (asset 3).

E is false.

The answer is: B, C, and D.

(Multiple correct AIT question) - Ch. 2.3 Reliability

Problem #9 Solution

Minimum mounting height ($min.\ MH$) in feet [ft] can be determined using the following formula[9]:

$$min.\ MH = \frac{max.\ candlepower}{1000} + ND$$

where:

Max. candlepower = The maximum luminous intensity (I) in candela [lm/sr].
ND = Mounting height (MH) constant.

The mounting height (MH) constant[10] for **long distribution** lighting is $ND = 15$.

Let's calculate the minimum mounting height (min. MH) using the 30,200 candela luminous intensity given in the problem:

$$\frac{30200}{1000} + 15 = \frac{226}{5} = 45.2$$

If your calculator gives you the answer as a fraction, like mine did above, you can convert from a fraction to decimal using the f↔d button.

The minimum mounting height is 45.2 feet. Normally, we would round down since a decimal of 0.2 is less than 0.5. However, since this value represents the **minimum** mounting height and the problem asks us to round to the nearest foot, we have to round up.

For example, 45.0 ft is **less than** the **minimum** mounting height of 45.2 feet and would be **incorrect**.

The minimum mounting height to the nearest foot is 46 feet.

The answer is: 46.

(Fill in the blank AIT question) - Ch. 2.4 Illumination, Lighting, and Energy Efficiency

---

[9] *NCEES® Reference Handbook (Version 1.1.2) - 2.2.3.2 Calculations (Illumination) p. 26*
[10] *NCEES® Reference Handbook (Version 1.1.2) - 2.2.3.1 Nomenclature (Illumination) p.24*

Problem #10 Solution

The **coincidence factor (F_C)**[11] is the ratio of the maximum demand of the group (D_g) to the sum (Σ) of the maximum demand of each individual in the group (D_i):

$$F_c = \frac{D_g}{\Sigma D_{i-max}}$$

```
                         DEG
   6.9
5+7.5+2.3
         0.466216216
```

The coincidence factor (F_C) expressed as a decimal is approximately 0.466.

Since the question asks for the coincidence factor (F_C) as a percentage, we can convert from a decimal to a percentage by multiplying by 100 (or by moving the decimal over to the right two places):

```
                         DEG
ans*100
          46.6216216
```

The percent coincidence factor (F_C) rounded to one decimal place as asked for in the problem is 46.6%.

The answer is: **46.6%**

(Fill in the blank AIT question) - Ch. 2.5 Demand Calculations

[11] *NCEES® Reference Handbook (Version 1.1.2) - 2.2.4 Demand Calculations p. 27*

Problem #11 Solution

Let's evaluate each of the possible answer choices:

(A) The amount of real power drawn by the motor from the power system is paid for at the energy cost rate.

Like transformers, motors have three main power attributes when looking at a power flow diagram. The power drawn by the motor from the power system (input power), the total power loss in the motor (power loss), and the power delivered to the load (output power).

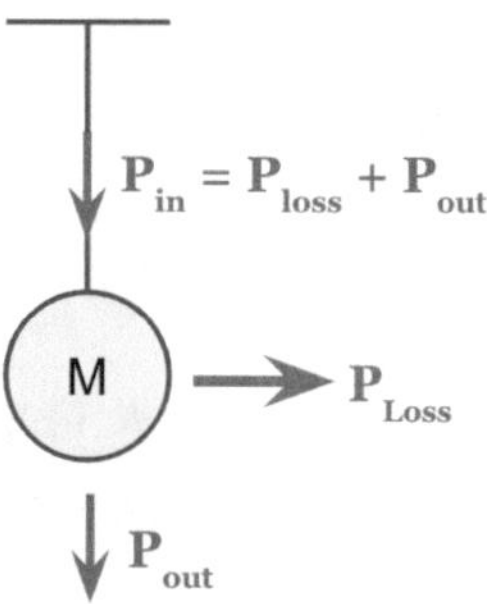

Neglecting any power factor penalties, we are only concerned with the real power quantities of all three in the unit of watts [W].

Since the real input power drawn by the motor (P_{in}) is equal to sum of the real power loss dissipated in the motor (P_{loss}) and the real output power (P_{out}) delivered to the load, the total amount of real power that is paid for is the real input power (P_{in}), which is the total amount of power provided by the power source.

A is true.

(B) The amount of apparent power drawn by the motor from the power system is paid for at the energy cost rate.

Apparent power $|S|$ is the magnitude of complex power in the unit of volt-amperes [VA]. Some utilities do charge customers that have poor power factor by the total volt-amperes [VA] instead of the total watts [W], but since the problem said to neglect any possible utility-imposed power factor penalty, this choice is false.

B is false.

(C) The amount of real power delivered by the motor to the load is paid for at the energy cost rate.

Looking at the motor power flow diagram above, notice that the total amount of power drawn from the power source (P_{in}) is equal to the sum of the power loss (P_{loss}) and the output power (P_{out}).

It's not just the energy delivered to the load (P_{out}) that costs money, but also any power that is lost along the way (P_{loss}) while delivering that power to the load.

C is false.

(D) Both the amount of real power delivered by the motor to the load and the total real power loss are paid for at the energy cost rate.

The sum of the output power (P_{out}) delivered by the motor to the load and the total power loss dissipated in the motor (P_{loss}) is equal to the input power drawn by the motor (P_{in}).

The input power drawn by the motor (P_{in}) is the total power that is provided by the power source and purchased by the customer.

D is true.

(E) All three are paid for at the energy cost rate: the amount of real power drawn by the motor from the power system, the amount of real power delivered by the motor to the load, and the total real power loss.

Since the input power (P_{in}) is equal to the sum of the output power (P_{out}) and power loss (P_{loss}), if we paid for all three quantities we would be paying for twice the amount of power that is supplied.

For example, consider a motor that is rated to deliver 75 kW of power (P_{out}) and has 10 kW of total power loss. The total input power drawn by the motor and supplied by the power source is:

$$P_{in} = P_{out} + P_{loss}$$
$$P_{in} = 75\ kW + 10\ kW$$
$$P_{in} = 85\ kW$$

The total amount of power supplied by the utility that the customer would be responsible for paying is $85\ kW$.

If the customer paid for the output power ($75\ kW$) and the power loss ($10\ kW$) on top of already paying for the total input power ($85\ kW$) drawn by the motor from the utility, the customer would end up paying for $170\ \mathrm{kW}$ of power when they were only supplied $85\ \mathrm{kW}$, a value that is twice as much.

E is false.

The answer is: **A and D.**

(Multiple correct AIT question) - Ch. 2.6 Energy Management

Problem #12 Solution

Let's evaluate each possible answer choice:

(A) The effective interest rate per month is 1%

Effective interest (i) is the interest per pay period[12]. Since the problem gives the interest charged per month, the given 1% interest is the effective monthly interest rate.

A is true.

(B) The nominal monthly interest rate is 1%

We can calculate the nominal (r) monthly interest rate with the following formula:

$$r = i \cdot m$$
$$r_{month} = i \cdot m_{month}$$
$$r_{month} = 1\%$$

Since the given effective interest rate (i) in the problem is the interest rate **per month**, the number of compounding periods (m) for the nominal **monthly** interest rate is equal to 1.

Another way to consider this is that there is only 1 month (m = 1) in a 1 month compounding period.

B is true.

(C) The nominal annual interest rate is 1%

We can calculate the nominal (r) annual interest rate with the same formula:

$$r = i \cdot m$$
$$r_{yr} = 1\% \cdot (12)$$
$$r_{yr} = 12\%$$

[12] *NCEES® Reference Handbook (Version 1.1.2) - 1.4.1 Nomenclature and Definitions (Eng. Econ.) p. 8*

The number of compounding periods (m) for the nominal annual interest rate is now 12 (m_{yr} = 12) since there are 12 months in one year.

Typically, the variable for nominal annual interest rate is just the letter r, but for the sake of clarity, we are using r_{yr}.

The nominal annual interest rate is 12%, not 1%.

C is false.

D) The nominal annual interest rate is 12%

In the previous step, we calculated that the nominal annual interest rate is 12%.

D is true.

(E) The effective annual interest rate is 12.7%

We can calculate the effective annual interest (i_e)[13] rate using the following formula:

$$i_e = \left(1 + \frac{r}{m}\right)^m - 1$$
$$i_e = \left(1 + \frac{12\%}{12}\right)^{12} - 1$$
$$i_e = 12.7\%$$

For the effective **annual** interest rate (i_e), m is the number of compounding periods **per year** (or per "annual"). There are 12 compounding periods per year since the given interest in the problem was per month.

Notice that we used the nominal annual interest rate calculated previously (r = 12%), but we could have used the interest rate per interest period period given in the problem (i = 1%) instead of r/m since i = r/m:

$$i_e = (1 + i)^m - 1$$
$$i_e = (1 + 1\%)^{12} - 1$$

[13] *NCEES® Reference Handbook (Version 1.1.2) - 1.4.2 Non-Annual Compounding p. 9*

$$i_e = 12.7\%$$

The annual effective interest rate (i_e) is approximately 12.7%.

E is true.

Let's look at a diagram to better visualize the different interest rates:

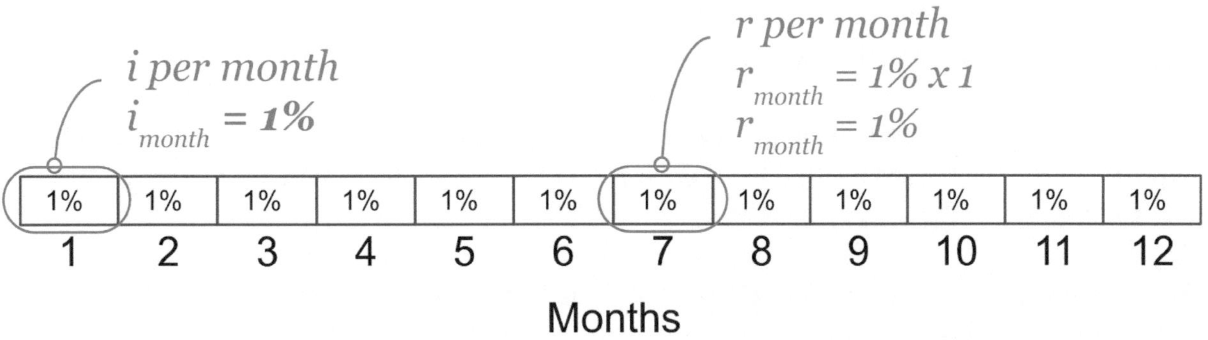

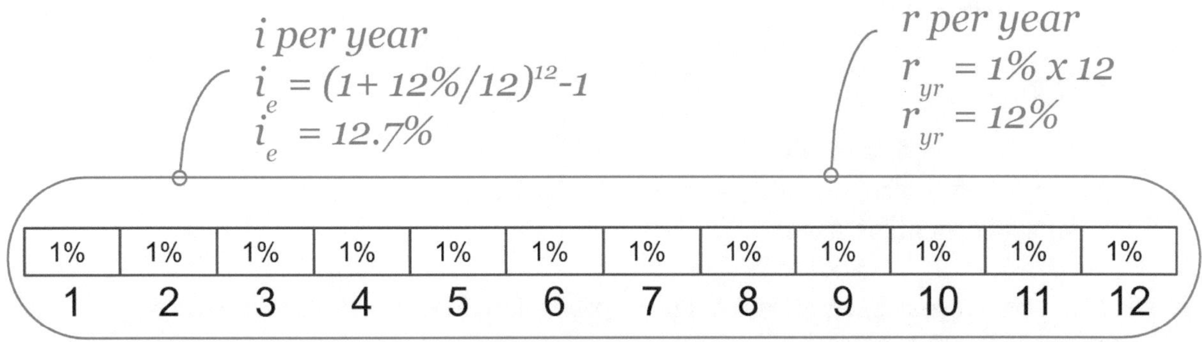

The answer is: **A, B, D, and E.**

(Multiple correct AIT question) - Ch. 2.7 Engineering Economics

Problem #13 Solution

Notice that every recovery period (years) listed in the ***modified accelerated cost recovery system* (MACRS)** table in the Reference Handbook[14] (3, 5, 7, and 10) has a recovery rate (percent) listed one year after the final recovery period year.

This is because MACRS assumes that assets are put into service halfway through the first year, and continue to depreciate in the following year after the final recovery period. This is known as the *half year convention*.

For example, the asset in the problem is a 5-year property class asset.

According to the MACRS table in the Reference Handbook, a 5-year property class asset will depreciate (recovery rate) by 20.00% in year 1, 32.00% in year 2, 19.20% in year 3, 11.52% in years 4 and 5, and 5.76% in year 6.

Even though it is a 5-year property class asset, due to the half year convention the final depreciation deduction will occur in year 6, one year after the recovery period.

We can calculate the depreciation deduction in any year for MACRS using the following formula:

$$D_j = (factor)C$$

where:

D_j = Year j depreciation deduction.
$factor$ = Year j recovery rate percent in the MACRS Factors table.
C = Initial cost of the asset.

The depreciation in year 6, which is the final depreciation deduction for the 5-year property class, for the \$55,000 asset is:

$$D_6 = \$55,000 \cdot (5.76\%)$$
$$D_6 = \$3,168$$

There are no decimals to round as the answer comes to a whole number.

The answer is: **$3,168**

(Fill in the blank AIT question) - Ch. 2.7 Engineering Economics

[14] *NCEES® Reference Handbook (Version 1.1.2) - 1.4.5.2 Modified Accelerated Cost Recovery System (MACRS) p. 9 - 10*

Problem #14 Solution

A solidly grounded wye power source provides a path back to the source for ground current to return through, which also allows ground-fault protection devices to monitor the current flowing through the ground. When a ground fault occurs, ground-fault protection will trip the service overcurrent protection devices.

False! **(A) Solidly grounded wye power source**

A high impedance grounded wye power source, like a solidly grounded wye power source, provides a path back to the source for ground current to return through, which also allows for ground-fault protection devices to monitor the current flowing through the ground.

However, the high impedance ground will also greatly limit the fault current according to Ohm's law ($I=V/Z$), which typically keeps the ground-fault current under the trip threshold except for the most extreme of ground faults. This allows the system to remain energized while allowing for ground-fault detection and troubleshooting.

True! **(B) High impedance grounded wye power source**

An ungrounded system will generally remain energized with just a single line-to-ground fault present on the system since there is no return path for ground-fault current to return to the source. With no return ground-fault path, there will also generally be no ground-fault monitoring protection devices.

True! **(C) Ungrounded wye power source**
True! **(D) Ungrounded delta power source**

The answer is: **B, C, and D.**

(Multiple correct AIT question) - Ch. 2.8 Grounding

Problem #15 Solution

Using a foot radius of 0.08 meters means we must use the surface layer derating factor (C_S) vs thickness of material (h_S) graph[15] for approximating the surface layer derating factor (C_S).

First, we'll need to calculate the reflection factor (K) so we can narrow down which of the curves in the graph to use.

The reflection factor (k) is calculated using the following formula[16]:

$$K = \frac{\rho - \rho_s}{\rho + \rho_s}$$

where:

ρ = The resistivity of the soil in the substation underneath any added resistive layer (like gravel or rock).
ρ_s = The resistivity of topsoil that is the direct walking surface, or the resistivity of any added resistive layer.

Let's calculate the reflection factor (k) using the 150 ohm meters of resistivity of the soil (ρ) and the 3 kiloohm meters of resistivity for the added layer of gravel (ρ_s):

```
                 DEG
150-3E3                 -19
150+3E3                  21
ans▸f◂▸d
           -0.904761905
```

The reflection factor (k) is approximately -0.9.

Out of the 10 total reflection factor (k) curves in the graph, we will only be using the one for k = -0.9.

Since the gravel depth is 0.1 meters, we'll draw a straight vertical line on the graph starting at the horizontal access at h_S = 0.1 meters until we reach the k = -0.9 curve.

Once we reach the curve, we'll draw a horizontal line left to determine the resulting value for the surface layer derating factor (C_S):

---

[15] *NCEES® Reference Handbook (Version 1.1.2) - 2.2.6.2 Step Voltage p. 28*
[16] *NCEES® Reference Handbook (Version 1.1.2) - 2.2.6.2 Step Voltage p. 28*

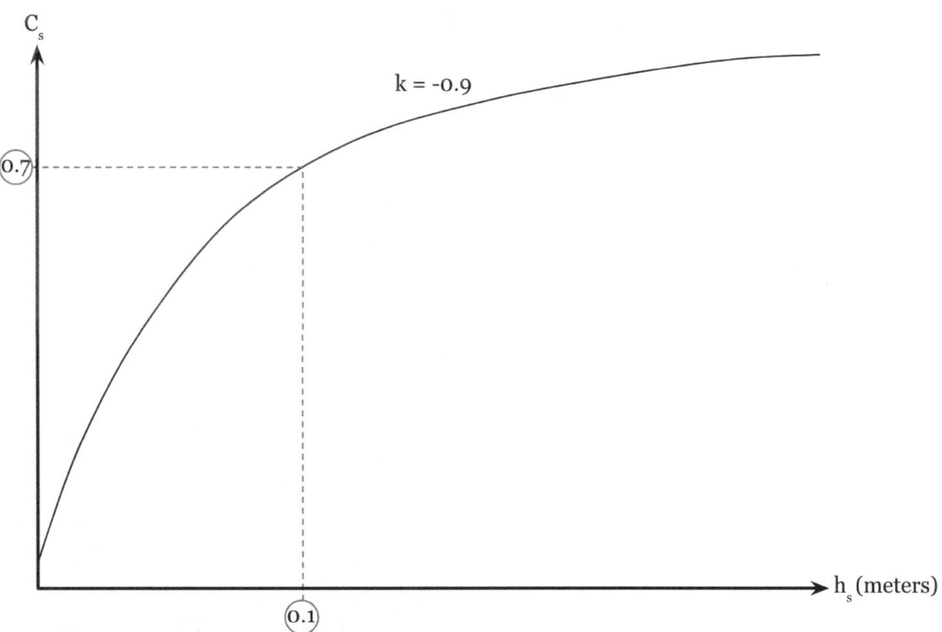

The resulting surface layer derating factor (C_S) is 0.7.

Even though the correct method to solve this problem is to use the graph because a foot radius of 0.08 meters was specified in the problem, we could have still arrived at the same answer using the following approximation formula[17] for the surface layer derating factor (C_S):

$$C_S = 1 - \frac{0.09\left(1 - \frac{\rho}{\rho_S}\right)}{2h_s + 0.09}$$

DEG

$$1-\frac{0.09\left(1-\frac{150}{3E3}\right)}{2(0.1)+0.09}$$

0.705172414

Rounded to one decimal place, the surface layer derating factor (C_S) is still 0.7.

The answer is: **0.7**

(Fill in the blank AIT question) - Ch. 2.8 Grounding

[17] *NCEES® Reference Handbook (Version 1.1.2) - 2.2.6.2 Step Voltage p. 28*

Problem #16 Solution

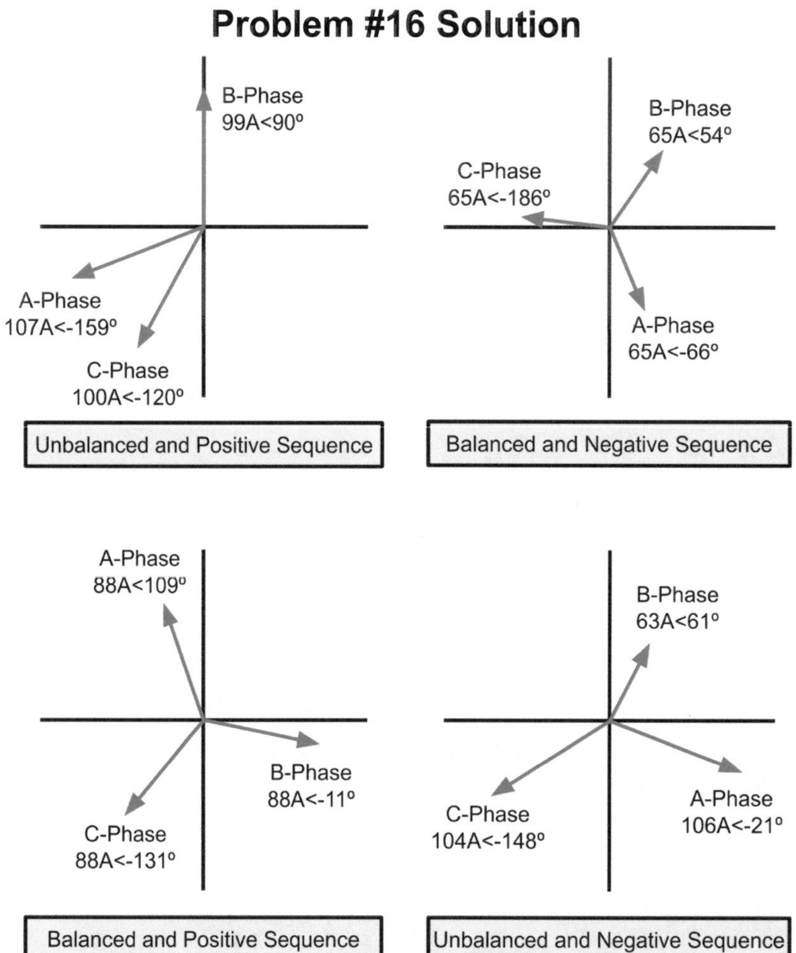

Balanced systems are both equal in magnitude and equal in phase angle displacement.

Unbalanced systems are either unequal in magnitude, unequal in phase angle displacement, or both.

For **positive (ABC) sequence** systems, A-phase leads B-phase, B-phase leads C-phase, and C-phase leads A-phase.

For **negative (ACB) sequence** systems, A-phase leads C-phase, C-phase leads B-phase, and B-phase leads A-phase.

In American and *IEEE®* standards, phasors always rotate in the **counterclockwise** direction regardless of phase sequence.

(Drag and drop AIT Question) - Ch. 4.4 Phasor Diagrams

Problem #17 Solution

Leading and lagging power factor relationships for synchronous machines based on their state of excitation are opposite when comparing generators to motors:

> **Over-excited** synchronous **generators** have a **lagging** power factor and **export** reactive power[18].
>
> **Under-excited** synchronous **generators** have a **leading** power factor and **import** reactive power.
>
> **Over-excited** synchronous **motors** have a **leading** power factor and **export** reactive power.
>
> **Under-excited** synchronous **motors** have a **lagging** power factor and **import** reactive power.

We can tell the difference between a synchronous generator and a synchronous motor phasor diagram by looking at the internal phase voltage (E_o) with respect to the terminal phase voltage (E):

> **Synchronous generators** have an internal phase voltage (E_o) that **leads** the terminal phase voltage (E).
>
> **Synchronous motors** have an internal phase voltage (E_o) that **lags** the terminal phase voltage (E).

We can tell if a synchronous machine (generator or motor) is over-excited or under-excited by comparing the real component of the internal phase voltage $Re\{E_o\}$ to the magnitude of the terminal phase voltage (E). This can be accomplished graphically by drawing a straight line from the internal phase voltage (E_o) to the horizontal axis as long as the terminal phase voltage (E) is at a reference of zero degrees:

> If the real component of the internal phase voltage $Re\{E_o\}$ is greater than the terminal phase voltage (E), the machine is **over-excited**:
>
> Over-excited: $Re\{E_o\} > E$

---

[18] *NCEES® Reference Handbook (Version 1.1.2) - 4.1.2 Equivalent Circuits and Characteristics p. 50-51*

If the real component of the internal phase voltage $Re\{E_o\}$ is less than the terminal phase voltage (E), the machine is **under-excited**:

Under-excited: $Re\{E_o\} < E$

Let's draw the real component of the internal phase voltage $Re\{E_o\}$ in red and evaluate each of the five possible choices:

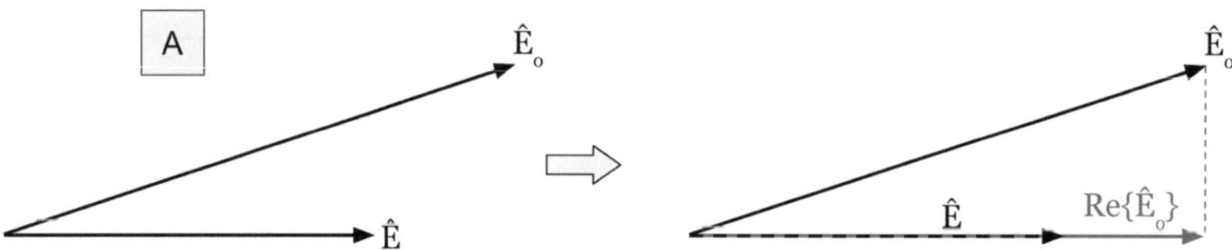

The internal phase voltage (E_o) **leads** the terminal phase voltage (E), so this is the phasor diagram for a **synchronous generator**.

The real component of the internal phase voltage $Re\{E_o\}$ is **greater than** the terminal phase voltage (E), so the machine is **over-excited.**

Synchronous **generators** that are **over-excited** supply reactive power and have a **lagging** power factor.

A is true.

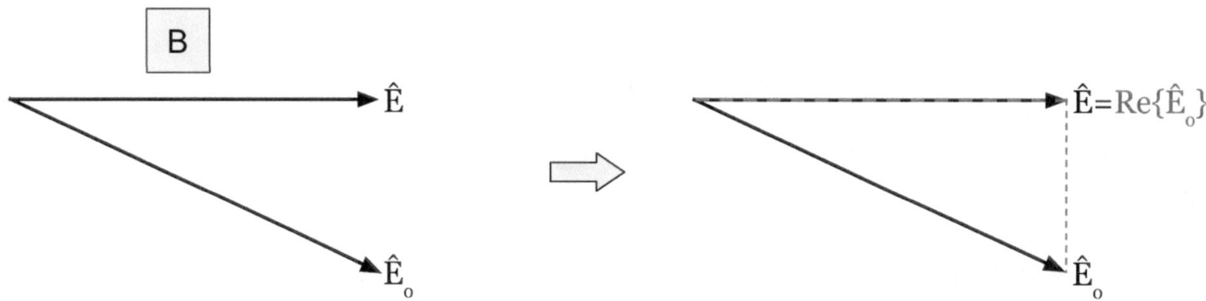

The internal phase voltage (E_o) **lags** the terminal phase voltage (E), so this is the phasor diagram for a **synchronous motor**.

The real component of the internal phase voltage $Re\{E_o\}$ is **equal to** the terminal phase voltage (E), so the machine is neither over- nor under-excited; it is in a state of **ideal excitation**.

Synchronous motors and generators that are ideally excited neither import nor export reactive power and have a **unity power factor** (PF = 1) that is neither leading or lagging.

B is false.

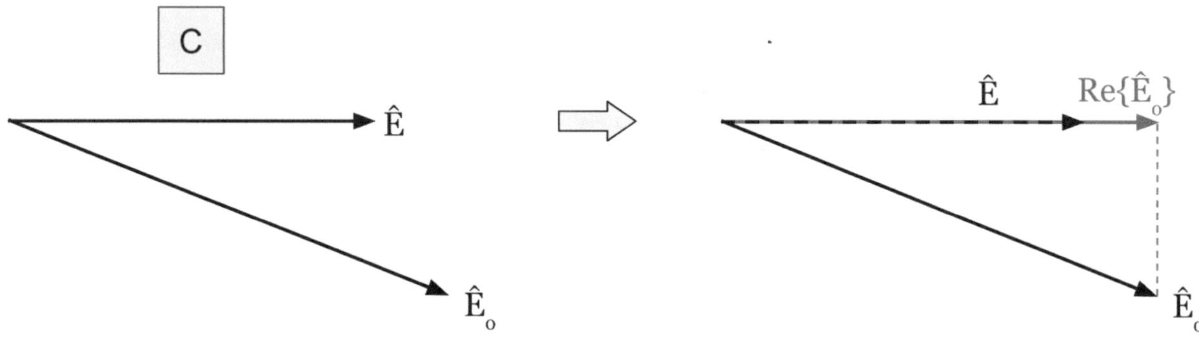

The internal phase voltage (E_o) **lags** the terminal phase voltage (E), so this is the phasor diagram for a **synchronous motor**.

The real component of the internal phase voltage $Re\{E_o\}$ is **greater than** the terminal phase voltage (E), so the machine is **over-excited.**

Synchronous **motors** that are **over-excited** export reactive power and have a **leading power factor**.

C is false.

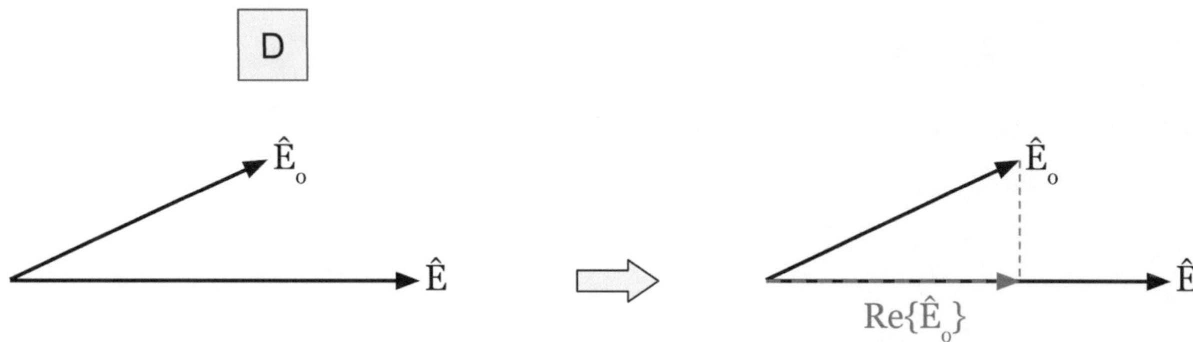

The internal phase voltage (E_o) **leads** the terminal phase voltage (E), so this is the phasor diagram for a **synchronous generator**.

The real component of the internal phase voltage $Re\{E_o\}$ is **less than** the terminal phase voltage (E), so the machine is **under-excited.**

Synchronous **generators** that are **under-excited** import reactive power and have a **leading power factor**.

D is false.

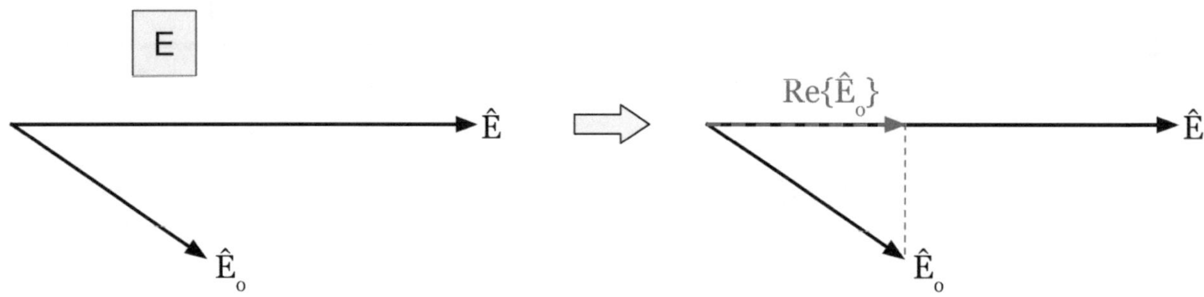

The internal phase voltage (E_o) **lags** the terminal phase voltage (E), so this is the phasor diagram for a **synchronous motor**.

The real component of the internal phase voltage $Re\{E_o\}$ is **less than** the terminal phase voltage (E), so the machine is **under-excited.**

Synchronous **motors** that are **under-excited** import reactive power and have a **lagging power factor**.

E is true.

The answer is: A and E.

(Multiple correct AIT question) - Ch. 4.4 Phasor Diagrams

Problem #18 Solution

The quickest way to solve for the line current delivered by each phase of the power supply is to convert the three-phase circuit to a single-phase equivalent circuit:

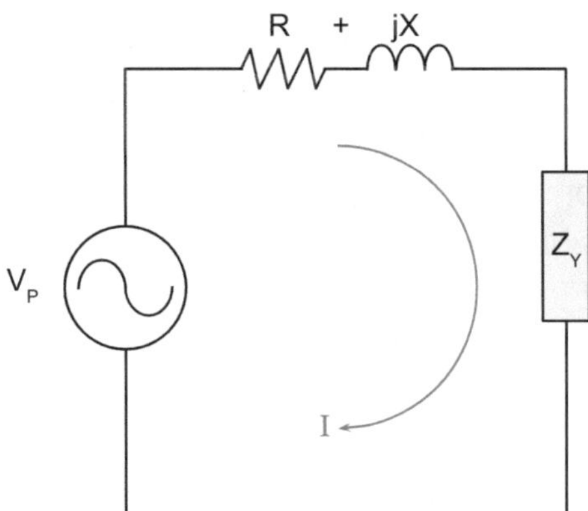

To use a single-phase equivalent circuit, first we'll need to convert each phase of the delta connected impedance (3,600 + j65 Ω) to the wye equivalent impedance per phase (Z_Y) by dividing by three:

$$\hat{Z}_Y = \frac{\hat{Z}_\Delta}{3}$$

```
DEG
3600+i65
   3
1200.19558∠1.0▸
```

Next, we'll use **Ohm's law** to solve for the current in the single-phase equivalent circuit (I) using the phase voltage of the system (V_P) and the sum of the total series complex impedance in the single-phase equivalent circuit ($Z_{eq} = R + jX + Z_Y$).

> ***Careful!*** *The phase voltage of the system (V_P) is the line-to-neutral voltage of the three-phase system line voltage, not the phase voltage of the delta connected load (this is a common mistake). This is true even if there is no neutral in the actual three-phase circuit we are analyzing like in this example.*

$$\hat{I} = \frac{\hat{V}_P}{\hat{Z}_{eq}}$$

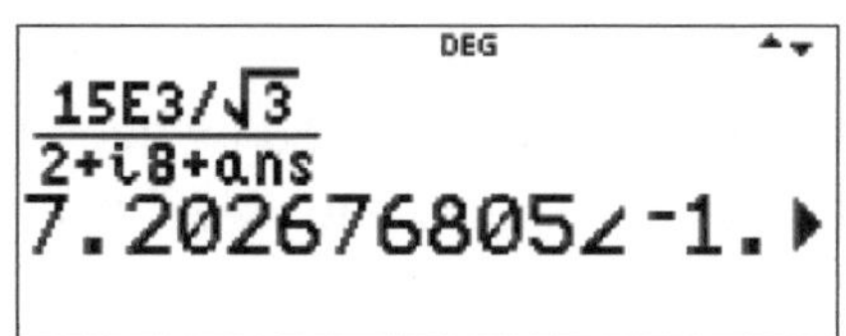

Since we are only asked for the magnitude of the current, any reference angle for the system phase voltage (V_P) may be used (try using different reference angles for V_P and notice that the magnitude of the current will not change, only the current phase angle will). To speed things up, I used zero degrees above by not including an angle for the phase voltage (V_P).

The magnitude of the current in the single-phase equivalent circuit rounded to the first decimal place is 7.2 amps.

The beauty of the single-phase equivalent circuit is that the current flowing in the single-phase equivalent circuit (I) will always equal the line current (I_L) flowing through each phase of a balanced three-phase circuit:

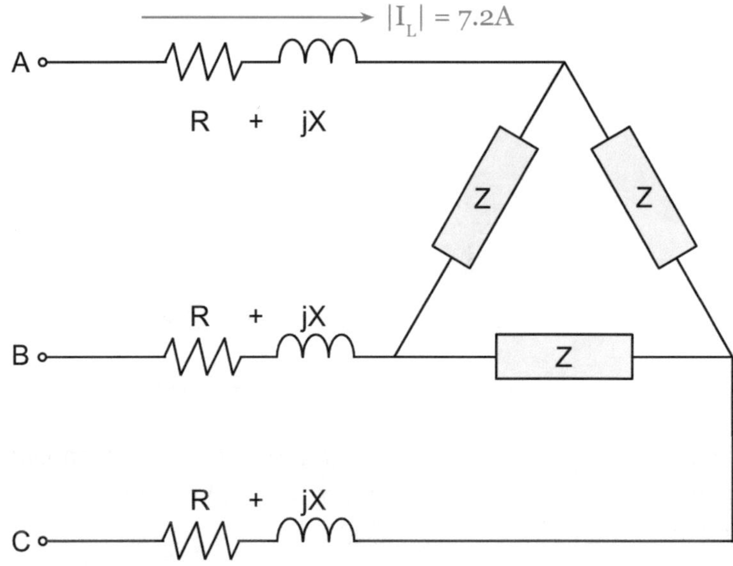

The magnitude of the line current delivered by each phase of the balanced and positive, three-phase, 60 Hz, 15 kV power supply rounded to the first decimal place is 7.2 A.

(Fill in the blank AIT question) - Ch. 4.1 Three-Phase Circuits

Problem #19 Solution

Let's draw the unbalanced delta connected load:

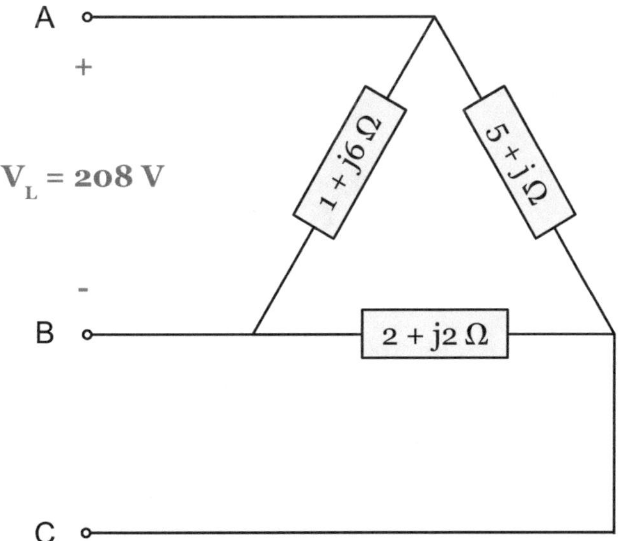

The balanced wye equivalent load will draw the same amount of three-phase power and have an equal impedance in each phase.

We can determine the total amount of three-phase power ($S_{3ø}$) drawn by the unbalanced delta connected load by summing up the power drawn by each individual phase:

$$\hat{S}_{3\phi} = \hat{S}_A + \hat{S}_B + \hat{S}_C$$

The power drawn by each phase can be calculated from the square of the phase voltage (Vp) to the conjugate of the phase impedance (Zp):

$$\hat{S}_{1\phi} = \frac{|V_p|^2}{\hat{Z}_p^*}$$

> *Note: this formula is not currently in the Reference Handbook but if you've attended our Live Class, this formula should be very familiar to you.*

Since the load is delta connected, the phase voltage (Vp) will be equal to the line voltage of the system (208V):

$$\hat{S}_{3\phi} = \frac{208V^2}{(1+j6\Omega)^*} + \frac{208V^2}{(2+j2\Omega)^*} + \frac{208V^2}{(5+j\Omega)^*}$$

$$\hat{S}_{3\phi} = 208V^2\left[\frac{1}{(1+j6\Omega)^*} + \frac{1}{(2+j2\Omega)^*} + \frac{1}{(5+j\Omega)^*}\right]$$

$$\hat{S}_{3\phi} = 208V^2\left(\frac{1}{1-j6\Omega} + \frac{1}{2-j2\Omega} + \frac{1}{5-j\Omega}\right)$$

DE DEG

$$208^2\left(\frac{1}{1-i6}+\frac{1}{2-i2}+\frac{1}{5-i}\right.$$

28149.434∠43.83

The unbalanced delta load draws approximately 28.1kVA<44° of three-phase complex power.

The balanced wye connected equivalent load will draw the same amount of three-phase power. However, since it is wye connected, the phase voltage (Vp) across each phase impedance will be equal to the line-to-neutral voltage:

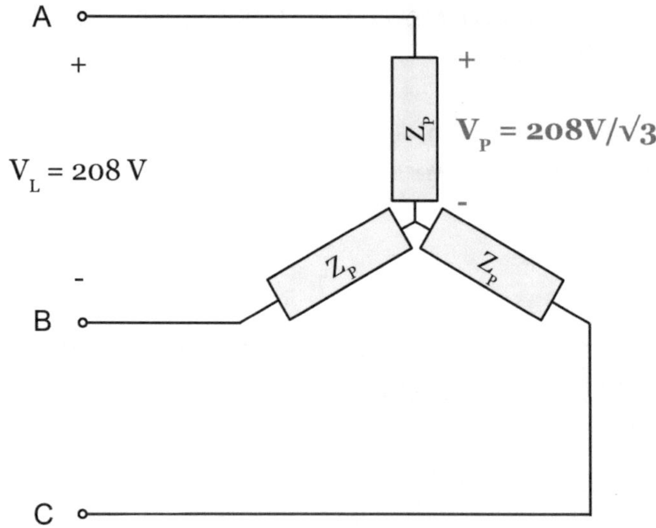

We can calculate the balanced wye equivalent impedance by using the same formula as before:

$$\hat{S}_{1\phi} = \frac{|V_p|^2}{\hat{Z}_p^*}$$

$$\hat{Z}_P = \frac{|V_p|^2}{\hat{S}_{1\phi}^*}$$

$$\hat{Z}_P = \frac{\hat{V}_p^2}{\hat{S}_{1\phi}^*}$$

$$\hat{Z}_P = \frac{\left(\frac{208V}{\sqrt{3}}\right)^2}{\left(\frac{28.1kVA<44^\circ}{3}\right)^*}$$

```
(208/√3)^2
conj(ans/3)
1.536940269∠43.
```

The complex impedance of the balanced wye equivalent load is approximately 1.5Ω<44° in polar form or 1.1 + j1.1 Ω in rectangular form.

The magnitude of the balanced wye equivalent load, rounded to one decimal place is 1.5 ohms.

The answer is: **1.5**

> *Note: since the problem only asks for the magnitude of the load, we could have simplified our calculations above using just magnitudes without having to worry about the conjugate (*):*
>
> $$|Z_P| = \frac{|V_P|^2}{|S_{1\phi}|}$$

```
(208/√3)^2
|ans/3|
1.536940269
```

> *The answer is still 1.5 ohms rounded to one decimal place.*

Alternative Solution:

If you're not comfortable with the $S = V^2/Z$ formula for power, another but more time consuming way to solve the problem is to first calculate each delta phase current using Ohm's law, followed by the complex power consumed by each phase.

I'm going to arbitrarily use a reference angle of zero degrees for the delta A-phase voltage, then lag each following delta phase voltage by -120 degrees for a balanced and positive system, but any voltage reference angle may be used and the math will still work out the same.

We can use the general formula of Ohm's law:

$$\hat{I}_P = \frac{\hat{V}_P}{\hat{Z}_P}$$

to calculate each of the delta phase currents:

$$\hat{I}_{AB} = \frac{\hat{V}_{AB}}{\hat{Z}_{AB}} \qquad \hat{I}_{BC} = \frac{\hat{V}_{BC}}{\hat{Z}_{BC}} \qquad \hat{I}_{CA} = \frac{\hat{V}_{CA}}{\hat{Z}_{CA}}$$

$$\hat{I}_{AB} = \frac{208V < 0^\circ}{1 + j6\,\Omega} \qquad \hat{I}_{BC} = \frac{208V < -120^\circ}{2 + j2\,\Omega} \qquad \hat{I}_{CA} = \frac{208V < 120^\circ}{5 + j\,\Omega}$$

$$\hat{I}_{AB} = 34.2A < -81^\circ \qquad \hat{I}_{BC} = 73.5A < -165^\circ \qquad \hat{I}_{CA} = 40.8A < 109^\circ$$

Next, we can use the general formula for complex single-phase power:

$$\hat{S}_{1\o} = \hat{V}_P \hat{I}^*_P$$

to calculate the complex power drawn by each phase of the unbalanced delta connected load:

$$\hat{S}_A = \hat{V}_{AB} \hat{I}^*_{AB}$$
$$\hat{S}_A = (208V < 0^\circ)(34.2A < -81^\circ)^*$$
$$\hat{S}_A = (208V < 0^\circ)(34.2A < 81^\circ)$$
$$\hat{S}_A = 7.1\,kVA < 81^\circ$$

$$\hat{S}_B = \hat{V}_{BC} \hat{I}^*_{BC}$$
$$\hat{S}_B = (208V < -120^\circ)(73.5A < -165^\circ)^*$$

$$\hat{S}_B = (208V < -120^\circ)(73.5A < 165^\circ)$$
$$\hat{S}_B = 15.3\ kVA < 45^\circ$$

$$\hat{S}_C = \hat{V}_{CA}\hat{I}^*_{CA}$$
$$\hat{S}_C = (208V < 120^\circ)(40.8A < 109^\circ)^*$$
$$\hat{S}_C = (208V < 120^\circ)(40.8A < -109^\circ)$$
$$\hat{S}_C = 8.5\ kVA < 11^\circ$$

We can now calculate the total three-phase complex power drawn by the unbalanced delta load by summing the complex power drawn by each phase:

$$\hat{S}_{3\phi} = \hat{S}_A + \hat{S}_B + \hat{S}_C$$
$$\hat{S}_{3\phi} = 7.1\ kVA < 81^\circ + 15.3\ kVA < 45^\circ + 8.5\ kVA < 11^\circ$$
$$\hat{S}_{3\phi} = 28.1\ kVA < 44^\circ$$

The unbalanced delta load draws approximately 28.1kVA<44º of three-phase complex power, the same value we arrived at in the original solution.

We can calculate the amount of complex power drawn by each phase of the balanced wye equivalent impedance by dividing this value by three:

$$\hat{S}_{1\phi} = \frac{\hat{S}_{3\phi}}{3}$$
$$\hat{S}_{1\phi} = \frac{28.1\ kVA < 44^\circ}{3}$$
$$\hat{S}_{1\phi} = 9.4\ kVA < 44^\circ$$

We can now use the same general formula for complex single-phase power:

$$\hat{S}_{1\emptyset} = \hat{V}_P\hat{I}^*_P$$

to calculate the phase current drawn by each phase of the balanced wye equivalent impedance:

$$\hat{S}_{1\emptyset} = \hat{V}_{AN}\hat{I}^*_{AN}$$

$$\hat{I}^*_{AN} = \frac{\hat{S}_{1ø}}{\hat{V}_{AN}}$$

$$\hat{I}^*_{AN} = \frac{9.4\,kVA < 44^\circ}{\frac{208V}{\sqrt{3}} < 0^\circ}$$

$$\hat{I}^*_{AN} = 78.1A < 44^\circ$$

$$\hat{I}_{AN} = 78.1A < -44^\circ$$

Last, we can use the general formula of Ohm's law once again:

$$\hat{Z}_P = \frac{\hat{V}_P}{\hat{I}_P}$$

to solve for the magnitude of the balanced wye equivalent impedance:

$$\hat{Z}_{AN} = \frac{\hat{V}_{AN}}{\hat{I}_{AN}}$$

$$\hat{Z}_{AN} = \frac{\frac{208V}{\sqrt{3}} < 0^\circ}{78.1A < -44^\circ}$$

$$\hat{Z}_{AN} = 1.5\,\Omega < 44^\circ$$

The complex impedance of the balanced wye equivalent load is approximately 1.5Ω<44° in polar form, or 1.1 + j1.1 Ω in rectangular form.

The magnitude of the balanced wye equivalent load, rounded to one decimal place is 1.5 ohms, the same value we arrived at in the first solution using the V^2/Z formula.

The answer is: **1.5**

> *Note:* all numerical values shown in the above alternative solution are rounded, but all decimal places were carried to the final value without rounding using the calculator.
>
> I used a reference angle of zero degrees for the wye equivalent A-phase line-to-neutral voltage (V_{AN}) to simplify the calculations. However, any reference angle would result in the same answer.

(Fill in the blank AIT question) - Ch. 4.1 Three-Phase Circuits

Problem #20 Solution

The variable designations for positive (1), negative (2), and zero sequence (0) were not included in the circuit diagrams in each answer choice to make this problem more challenging. We've included these variables in the diagrams below to serve as a learning aid.

Let's work through each of the circuit diagrams in the order that we would typically draw them.

First is the **single line diagram**:

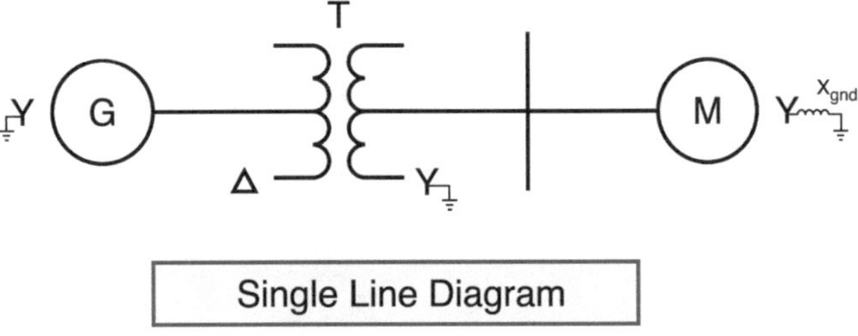

Single Line Diagram

Single line diagrams are used to represent three-phase systems by showing only one phase since all phases are assumed to be balanced.

Next, we can draw the **positive sequence network** by drawing all electric and magnetic field (EMF) sources such as generators and motors as a voltage source in series with the machine's impedance, drawing all remaining devices such as transformers and line impedances either in series or parallel with respect to how they are connected in the single line diagram, and connecting a reference bus back to the source:

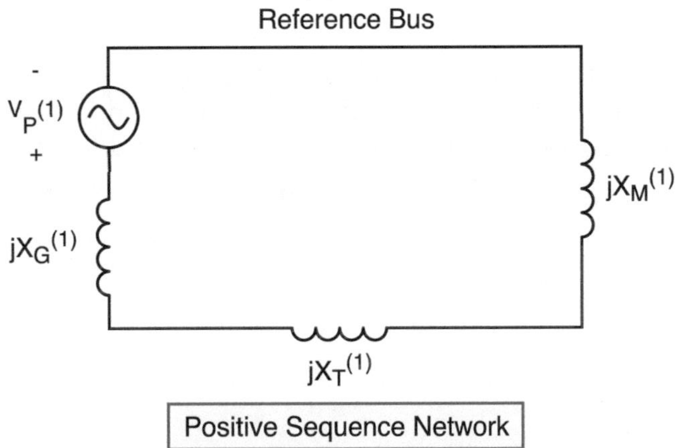

Positive Sequence Network

The reference bus for positive, negative, and zero sequence circuits can be shown either at the top or the bottom of the circuit.

Only the positive sequence impedance ($Z^{(1)}$) of each device is used in the positive sequence network.

It is typical to combine every voltage source present in the positive sequence network into one equivalent voltage source just as we have done here with the voltage source for the generator and motor.

It is common to only use reactance (X) values since they are much larger compared to resistive values (R) for mostly inductive machines like transformers, motors, and generators unless a much more in-depth analysis is being done that also includes the resistance values of each to form the complex impedance of each circuit element.

Next, we can draw the **negative sequence network** by short circuiting all voltage sources and changing all impedance values from each device's positive sequence impedance ($Z^{(1)}$), to their negative sequence impedance ($Z^{(2)}$):

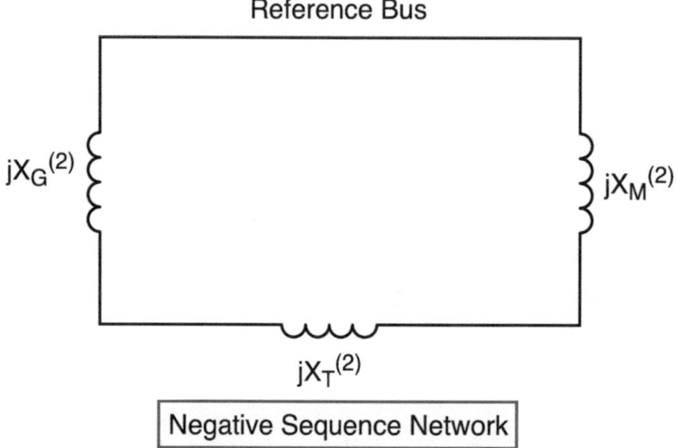

Last, we can draw the **zero sequence network** by changing each transformer connection to the reference bus depending on the transformer connection type[19], changing each generator and motor connection to the reference bus depending on its grounding configuration[20], and changing all impedance values from each device's negative sequence impedance ($Z^{(2)}$) to their zero sequence impedance ($Z^{(0)}$):

[19] *NCEES® Reference Handbook (Version 1.1.2) - 5.1.5 Fault Current Analysis p. 70*
[20] *NCEES® Reference Handbook (Version 1.1.2) - 5.1.5 Fault Current Analysis p. 69*

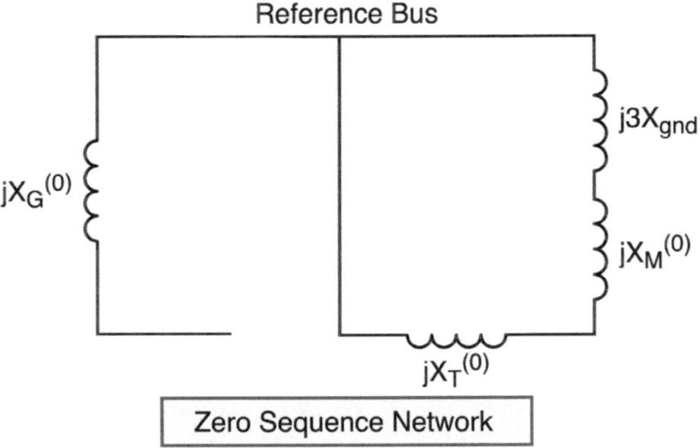

jX_{gnd} represents the grounding reactance of the motor and $j3X_{gnd}$ shown above (the motor grounding reactance multiplied by three) represents the value of this reactance in the zero sequence network.

> *Note: grounding impedance is only present on the* ***zero sequence network*** *and is* ***always*** *multiplied by three.*

Completed drag and drop diagram:

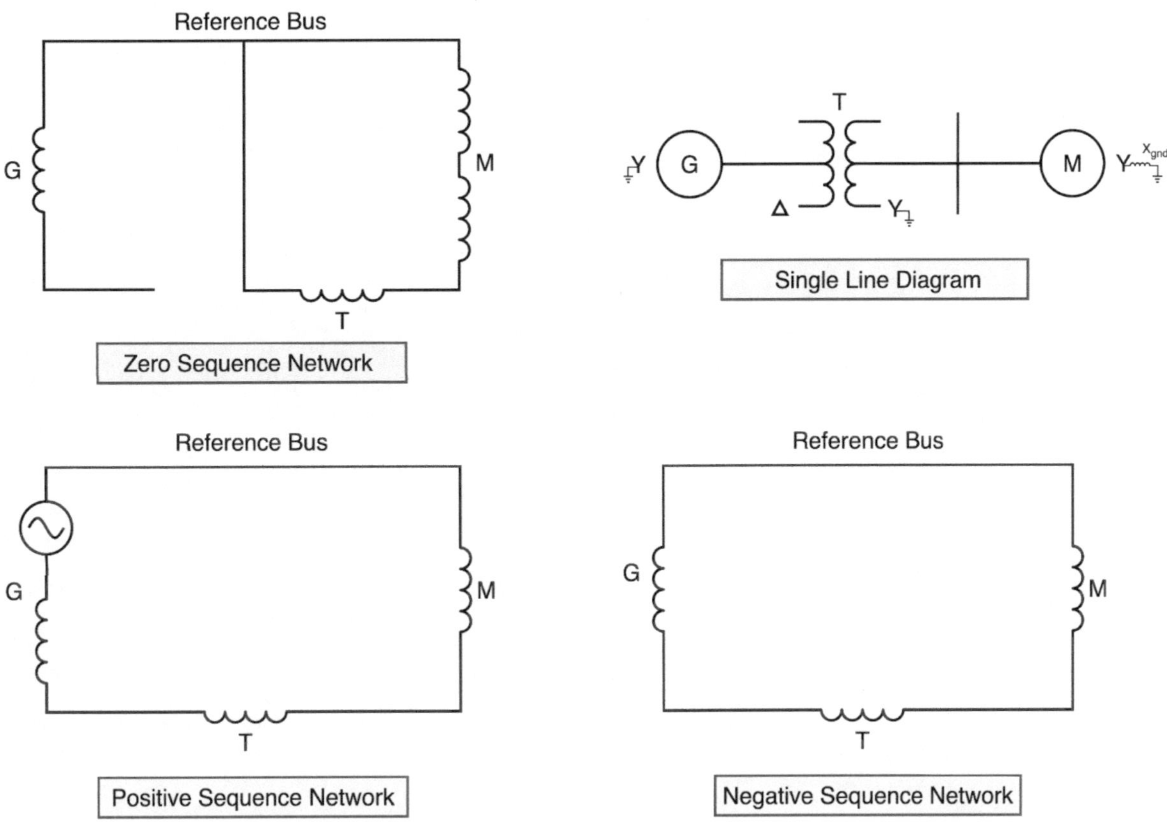

Note that the diagram in the problem did not distinguish between the positive (1) negative (2) and zero (0) sequence reactances in order to make the problem more difficult.

The answer is: **see diagram above.**

(Drag and drop AIT question) - Ch. 4.2 Symmetrical Components

Problem #21 Solution

First, let's start by drawing the zero sequence network of the system:

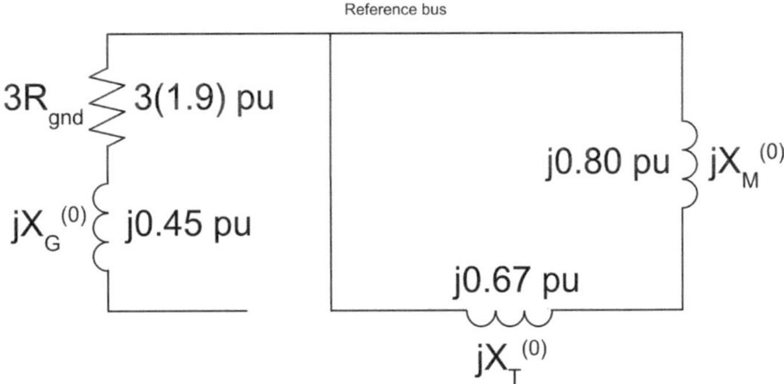

Next, we'll need to show the fault location to see how each of the impedances in the zero sequence network add up:

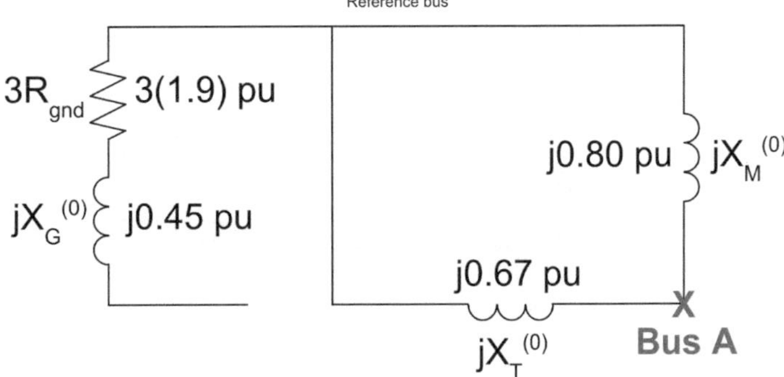

With a fault on bus A:

- The generator grounding resistance ($3R_{gnd}$) and the generator zero sequence reactance ($X_G^{(0)}$) are open circuited on the zero sequence network by the primary delta connection of the transformer.
- The transformer zero sequence reactance ($X_T^{(0)}$) is now in parallel with the motor zero sequence reactance ($X_M^{(0)}$).

The last step is to calculate the total equivalent complex impedance of the zero sequence network ($Z_{eq}^{(0)}$) by adding the transformer zero sequence reactance ($X_T^{(0)}$) in parallel with the motor zero sequence reactance ($X_M^{(0)}$):

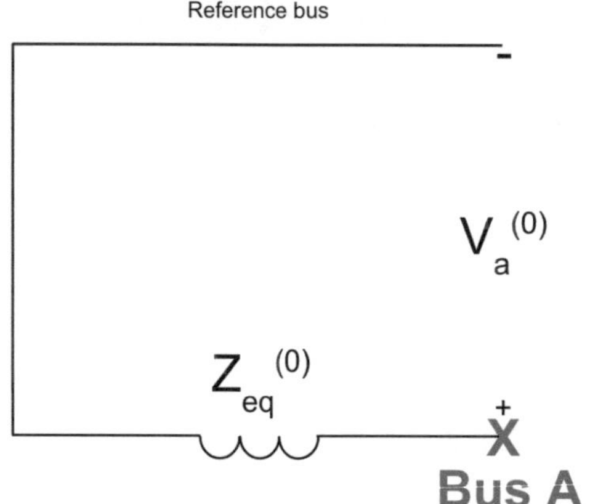

$$\hat{Z}_{eq}^{(0)} = jX_T^{(0)} // jX_M^{(0)}$$

$$\hat{Z}_{eq}^{(0)} = \frac{jX_T^{(0)} \cdot jX_M^{(0)}}{jX_T^{(0)} + jX_M^{(0)}}$$

$$\hat{Z}_{eq}^{(0)} = \frac{(j0.67\,pu)(j0.80\,pu)}{j0.67\,pu + j0.80\,pu}$$

```
DEG
i0.67*i0.80
i0.67+i0.80
        0.36462585i
```

The total equivalent complex impedance of the zero sequence network ($Z_{eq}^{(o)}$) rounded to two decimal places is 0 + j0.36 pu.

The answer is: **0 + j0.36**

(Fill in the blank AIT question) - Ch. 4.2 Symmetrical Components

Problem #22 Solution

When working with the per unit system, each transformer divides the system into separate voltage zones. For example, consider the three voltage-zone system created by the transformers in the hypothetical one line diagram below:

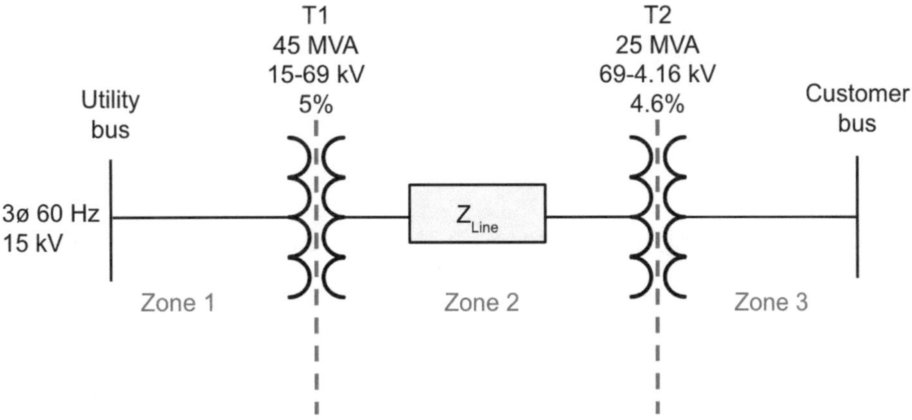

The new base power and base voltage are selected in one particular voltage zone. For example, we will arbitrarily select a power base of 50 MVA and a voltage base of 4.16 kV for zone 3:

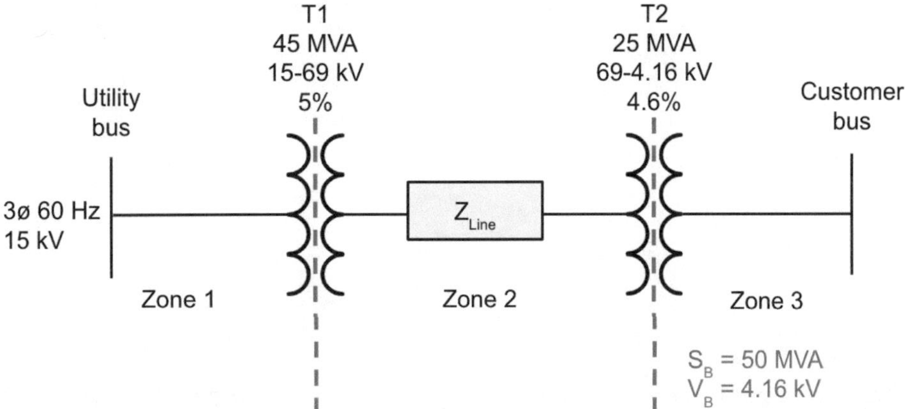

When working with the per unit system, transformers are typically assumed to be ideal so that the power flowing into each transformer is identical to the power flowing out of that transformer. Because of this, the base power for each remaining zone will be identical:

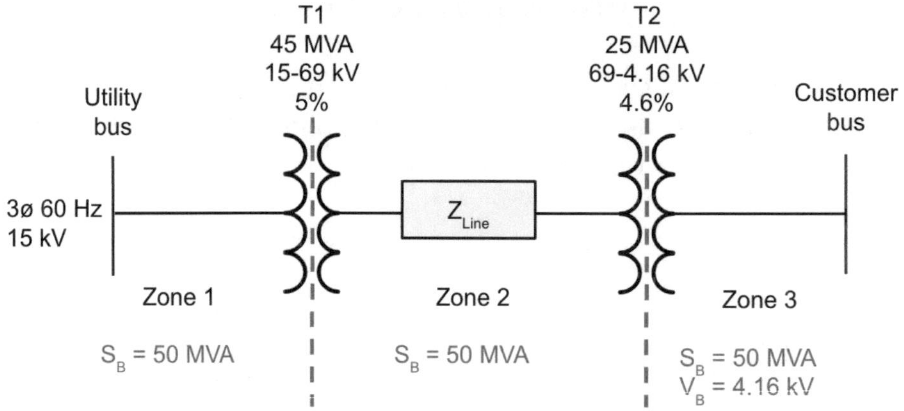

We can calculate the remaining base voltages in each zone using the transformer ratio formula[21]:

$$a = \frac{N_1}{N_2} = \frac{V_1}{V_2}$$

Since we are starting at the very most downstream zone, we can work our way up by solving for the primary voltage (V_1) of each transformer.

The base voltage in zone 2, using the base voltage in zone 3 as the voltage on the secondary of T2 (V_2), is:

$$V_1 = V_2\left(\frac{N_1}{N_2}\right)$$

$$V_1 = 4.16\ kV\left(\frac{69\ kV}{4.16\ kV}\right)$$

$$V_1 = 69\ kV$$

The base voltage in zone 1, using the base voltage in zone 2 as the voltage on the secondary of T1 (V_2), is:

$$V_1 = V_2\left(\frac{N_1}{N_2}\right)$$

$$V_1 = 69\ kV\left(\frac{15\ kV}{69\ kV}\right)$$

$$V_1 = 15\ kV$$

[21] *NCEES® Reference Handbook (Version 1.1.2) - 4.3.1.1 Turns ratio in an ideal transformer p. 58*

Here is the completed diagram showing the base power and base voltage in each zone:

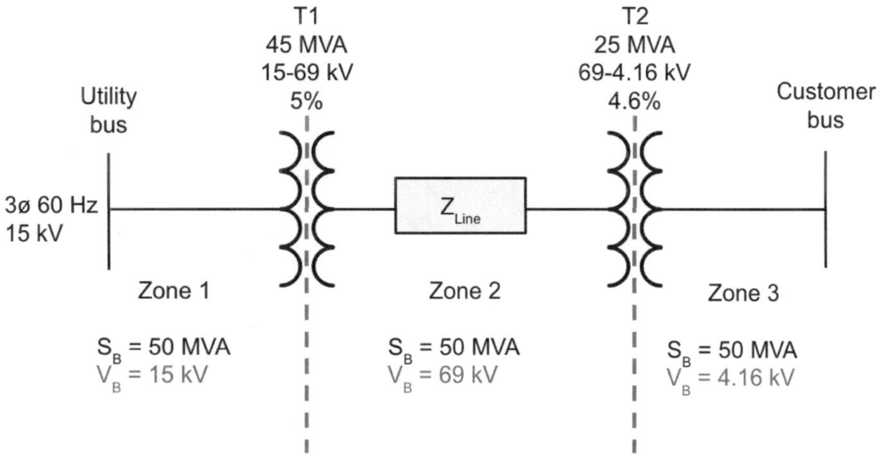

Notice that the base power in every voltage zone is equal, but the base voltage depends on each transformer ratio.

We can use this information to evaluate answer choices (A) and (B):

> False! **(A) The system base power and base voltage are the same in every voltage zone.**
>
> True! **(B) The system base power is the same in every voltage zone; the voltage base changes depending on each transformer ratio.**

The base impedance in each zone is calculated from the base power and base voltage in each specific zone using the base impedance formula[22]:

$$Z_B = \frac{V_B^2}{S_B}$$

Since the base voltage (V_B) in each zone is different, the base impedance in each zone will not be equal.

For example, the base impedance in zone 3 is:

[22] *NCEES® Reference Handbook (Version 1.1.2) - 3.1.3 Base Impedance p. 34*

$$Z_B = \frac{4.16\ kV^2}{50\ MVA}$$

$$Z_B = 0.346\ \Omega$$

The base impedance in zone 2 is:

$$Z_B = \frac{69\ kV^2}{50\ MVA}$$

$$Z_B = 95.22\ \Omega$$

The base impedance in zone 1 is:

$$Z_B = \frac{15\ kV^2}{50\ MVA}$$

$$Z_B = 4.5\ \Omega$$

Here is the completed diagram showing the base impedance in each zone:

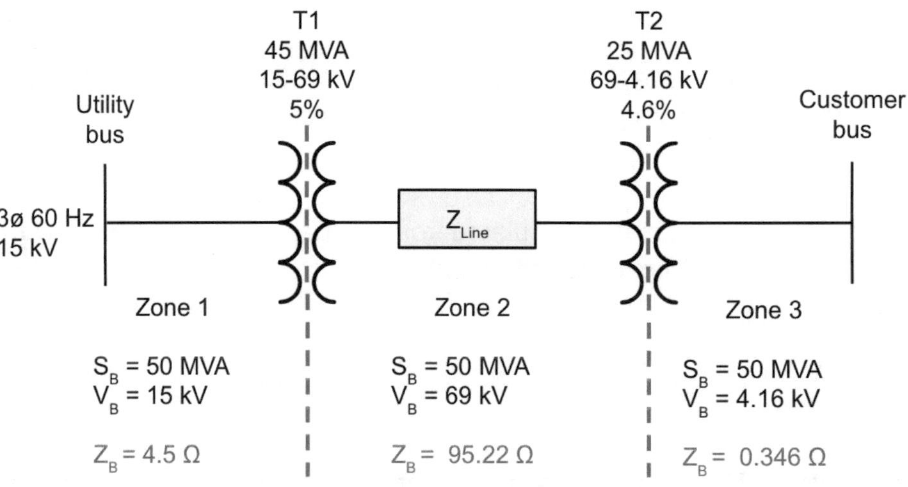

We can use this information to evaluate answer choice (C):

True! **(C) The system base impedance and base voltage in each voltage zone change depending on each transformer ratio.**

When the actual voltage and the base voltage are equal, a special condition occurs. For example, in zone 1 the three-phase line voltage at the utility bus (V) and the base voltage (V_B) in that zone are both equal to 15 kV.

This means that the per unit voltage (V_{pu}) in zone 1 is equal to 1 pu:

$$V_{pu} = \frac{V}{V_B}$$
$$V_{pu} = \frac{15\ kV}{15\ kV}$$
$$V_{pu} = 1\ pu$$

Let's substitute this into the per unit apparent power formula (S_{pu}):

$$S_{pu} = V_{pu} \cdot I_{pu}$$

$$V_{pu} = 1\ pu$$

$$S_{pu} = 1\ pu \cdot I_{pu}$$
$$S_{pu} = I_{pu}$$

Looking at the relationship above, the per unit apparent power (S_{pu}) is equal to the per unit current (I_{pu}).

Had we used the per unit system for fault current analysis and solved for the short circuit fault duty (MVA) in per unit (S_{pu}), it would be equal to the short circuit per unit current (I_{pu}):

> True! **(D) When the voltage is equal to the base voltage, the short circuit per unit fault duty is equal to the short circuit per unit current.**

Last, let's look at this same special condition and see how it relates to impedance by substituting it into Ohm's law:

$$V_{pu} = I_{pu} \cdot Z_{pu}$$

$$V_{pu} = 1\ pu$$

$$1\ pu = I_{pu} \cdot Z_{pu}$$

$$I_{pu} = \frac{1}{Z_{pu}}$$

Looking at the relationship above, the per unit current (I_{pu}) is inversely proportional to the total equivalent per unit impedance (Z_{pu}).

Had we used the per unit system for fault current analysis and solved for the short circuit per unit current (I_{pu}), it would be inversely proportional to the total equivalent per unit impedance (Z_{pu}).

> True! **(E) When the voltage is equal to the base voltage, the short circuit per unit current is inversely proportional to the total equivalent per unit impedance.**

The answer is: **B, C, D, and E.**

(Multiple correct AIT question) - Ch. 4.3 Per unit system

> Note: *the per unit system can be confusing at first. If you have not already, be sure to work through the free per unit example article that is available to everyone on our main webpage:*
>
> Per Unit Example – How To, Tips, Tricks, and What to Watch Out for on the Electrical PE Exam:
> https://www.electricalpereview.com/per-unit-example-tips-tricks-watch-electrical-pe-exam/

Problem #23 Solution

In RC and RL transient DC circuits[23], steady state conditions occur at a time equal to 5 times the time constant (τ). This relationship is **not** mentioned in the Handbook and should be remembered:

$$Steady\ state\ occurs\ at\ =\ 5\tau$$

For an **RC circuit**, the time constant (τ) is equal to the product of the resistance (R) and capacitance (C):

$$\tau = RC$$

For an **RL circuit**, the time constant (τ) is equal to the ratio of inductance (L) to resistance (R):

$$\tau = \frac{L}{R}$$

To find the time when the voltage across the capacitor in the RC circuit in the problem will reach steady state conditions, all we need to do is multiply the time constant (τ) by 5:

$$5\tau = 5RC$$
$$5\tau = 5(9.5\ k\Omega)(6\ \mu F)$$

```
DEG
5(9.5E3)(6E-6)
          0.285
```

The voltage across the capacitor in the RC circuit will reach steady state conditions at 0.285 seconds, or 285 milliseconds.

The answer is: **285**

(Fill in the blank AIT question) - Ch. 4.6 DC Circuits

[23] *NCEES® Reference Handbook (Version 1.1.2) - 3.1.5 DC Circuits p. 37*

Problem #24 Solution

Since this is a series circuit, we can calculate the current flowing through each element in the circuit, including the capacitor, at any time when or after the switch is closed ($t \geq 0$) using the following formula for an RC transient circuit[24] (the following formulas in this solution will be enlarged since the exponents are small and hard to see):

$$i(t) = \left(\frac{V - v_c(0)}{R}\right)e^{-\frac{t}{RC}}$$

where:

$i(t)$ = The current flowing in the series RC circuit.
V = The DC voltage source in volts.
$v_c(t)$ = The initial voltage across the capacitor before the switch is closed.
t = The variable time in seconds.
R = The total series resistance in the circuit in ohms [Ω].
C = The capacitance of the capacitor in farads [F].

First, let's set up the formula for the current flowing in the series RC circuit using the values given in the problem:

$$i(t) = \left(\frac{V - v_c(0)}{R}\right)e^{-\frac{t}{RC}}$$

$$i(t) = \left(\frac{35\,V - 14\,V}{7.5\,k\Omega}\right)e^{-\frac{t}{(7.5\,k\Omega)(4\,\mu F)}}$$

$$i(t) = 2.8\,mA \cdot e^{-\frac{t}{0.03\,s}}$$

Next, let's plug in t = 60 ms and use our calculator to solve for the current at 60 milliseconds (60 ms = 0.06 seconds):

[24] *NCEES® Reference Handbook (Version 1.1.2) - 3.1.5 DC Circuits p. 37*

$$i(t) = 2.8\,mA \cdot e^{-\frac{0.06\,s}{0.03\,s}}$$

```
DEG
2.8E-3e^-0.06/0.03
        0.000378939
```

The current flowing through the capacitor at t = 60 milliseconds is 0.00038 amperes, or 0.4 milliamps rounded to one decimal place.

The answer is: **0.4**

(Fill in the blank AIT question) - Ch. 4.6 DC Circuits

For extra practice, let's take this one step further and graph the current flowing in the series RC circuit with respect to time so that we have a better understanding of what we just solved for, and how this value behaves with time in order to be better prepared for qualitative questions on the PE exam:

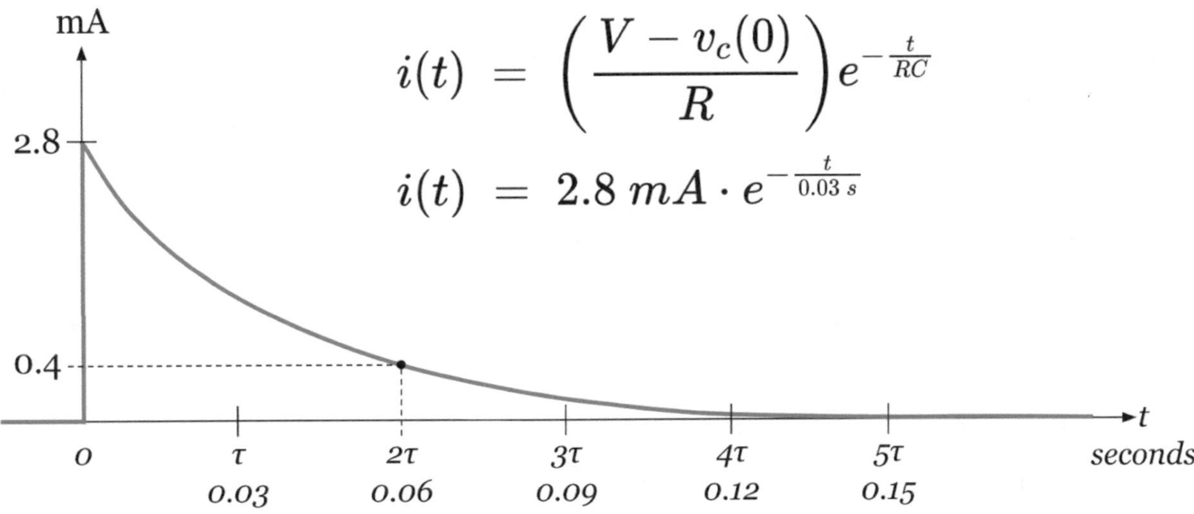

Notice that prior to the switch being closed (t < 0), there is no current flowing through the circuit. With the switch open, this is an open circuit and no current can flow.

Notice that right at the moment that the switch closes (t = 0), the current flowing in the circuit is equal to 2.8 mA. This is the current drawn by the series resistor (R) and capacitor (C) as the capacitor begins to charge.

Notice that during the **transient period** from t = 0 to t = 5τ, the current rapidly decays exponentially until it approaches **steady state conditions** at t = 5τ.

At **steady state conditions,** the current in a series DC RC circuit will always equal zero amps as the capacitor becomes fully charged, and it will remain zero until the switch is opened again and the capacitor can discharge.

Last, notice the current marked on the graph that we solved for in this problem at t = 0.06 seconds (or 60 milliseconds); this just so happens to be equal to t = 2τ for this problem.

For extra practice, try solving for the current at t = 5τ. It should be approximately zero as should any value for time greater than 5τ.

Problem #25 Solution

We can calculate the voltage across the inductor (L) in a DC series RL circuit at any time when or after the switch is closed (t ≥ 0) using the following formula[25]:

$$v_L(t) \ = -i(0)Re^{-\frac{Rt}{L}} + Ve^{-\frac{Rt}{L}}$$

where:

$v_L(t)$ = The voltage across the inductor.
$-i(0)$ = The initial current flowing through the inductor before the switch is closed.
R = The total series resistance in the circuit in ohms [Ω].
t = The variable time in seconds.
L = The inductance of the inductor in henries [H].
V = The DC voltage source in volts.

Since there is no current flowing through the inductor prior to the switch closing, $i(0) = 0$. Let's substitute this back into the formula above to simplify, then plug in all remaining values given in the problem:

$$\begin{aligned} v_L(t) \ &= -i(0)Re^{-\frac{Rt}{L}} + Ve^{-\frac{Rt}{L}} \\ v_L(t) \ &= -(0)Re^{-\frac{Rt}{L}} + Ve^{-\frac{Rt}{L}} \\ v_L(t) \ &= Ve^{-\frac{Rt}{L}} \\ v_L(t) \ &= 45 \cdot e^{-\frac{1\,k\Omega \cdot t}{2\,H}} \\ v_L(t) \ &= 45 \cdot e^{-500t} \end{aligned}$$

Before we answer this question, let's graph the voltage across the inductor in the series RL circuit with respect to time so that we have a better understanding of what we just solved for, and how this value behaves with time in order to be better prepared for qualitative questions on the PE exam:

[25] *NCEES® Reference Handbook (Version 1.1.2) - 3.1.5 DC Circuits p. 37*

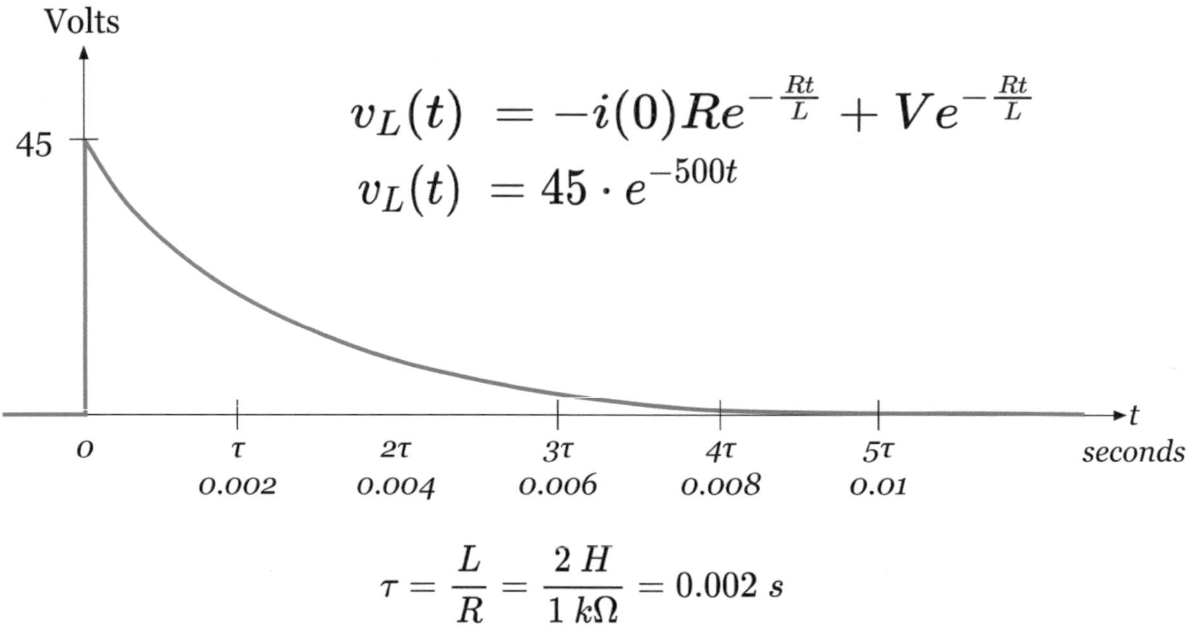

$$\tau = \frac{L}{R} = \frac{2\ H}{1\ k\Omega} = 0.002\ s$$

Notice that prior to the switch closing, the voltage across the inductor is zero volts.

At the exact moment that the switch is closed, the voltage across the inductor is equal to 45 volts and then rapidly decays exponentially until reaching a steady state voltage of zero volts at a time approximately equal to 5 times the time constant (τ).

> *Note: it's helpful to notice that if we already had enough familiarity with RL circuits, we could have easily solved this problem quickly without using any math by recognizing that the voltage across the inductor would be equal to the input DC voltage (45 volts in this problem) due to the fact that the initial current prior to the switch closing is equal to zero and that this term would cancel in the inductor voltage formula.*

The voltage across the inductor at the exact moment that the switch closes is 45 volts.

The answer is: **45**

(Fill in the blank AIT question) - Ch. 4.6 DC Circuits

Problem #26 Solution

Using the official *NCEES® Reference Handbook*, there are three different types of battery potential (voltage) values available, it will be up to us to choose the correct value based on context clues from the problem.

The first two options are either the **theoretical voltage value (V)** or **practical battery nominal voltage (V)** from the *Voltage, Capacity, and Specific Energy of Major Battery Systems table*[26].

The third option is to calculate the **standard potential voltage**[27] of the cell using the formula by the same name and the anode and cathode **standard reduction potential voltage values at 25° C** from the *Characteristics of Typical Electrode Materials* table[28].

For these types of problems, you'll need to use the context clues from the problem to determine which value is being asked for.

> **Theoretical voltage** includes only the active anode and cathode electrode and ignores other material that may be involved in the chemical reaction taking place in the cell such as the electrolyte solution, water, and evaporation. See table 3.2.1.1. note "a" for more details.
>
> **Practical (nominal) voltage** is what you would expect to see in real use cases based on "real world" operating conditions. Think of practical values as "off the shelf" characteristics for each cell. See table 3.2.1.1. note "b" for more details.
>
> **Standard voltage** includes every aspect of the chemical reaction in a cell such as the electrolyte solution, water, evaporation, in addition to the active anode and cathode electrode materials.

Since the problem is asking for the most likely every day voltage of each cell that can be expected, the correct value is the **practical voltage** of the cell.

Practical voltage can be found in table 3.2.1.1.

For a lithium-ion battery (battery type), the practical battery voltage is 3.8 volts.

The answer is: 3.8

---

26 *NCEES® Reference Handbook (Version 1.1.2) - 3.2.1.1 Voltage, Capacity, and Specific Energy of Major Battery Systems—Theoretical and Practical Values p. 39-40*

27 *NCEES® Reference Handbook (Version 1.1.2) - 3.2.1.2 Standard Potential of a Cell p. 41*

28 *NCEES® Reference Handbook (Version 1.1.2) - 3.2.1.4 Characteristics of Typical Electrode Materials p. 42*

(Fill in the blank AIT question) - Ch. 5.1 Battery Characteristics and Ratings

Problem #27 Solution

To answer this question correctly, you'll need to reference the *Relationship of Temperature on Battery Capacity*[29] graph in the *NCEES® Reference Handbook*.

Looking at the graph, the DC amps delivered by a battery to a DC load is represented by the horizontal axis labeled *discharge current*. As each curve on the graph increases to the right along the horizontal axis, the amount of DC amps delivered by a battery increases.

The percent of total rated amp hours (capacity) a battery is able to deliver is represented by the vertical axis labeled *percent of rated capacity*. As each curve decreases downward along the vertical axis, the amount of total rated amp hours a battery is able to deliver decreases.

There are a total of six curves on the graph labeled T1 through T6. Each curve represents battery capacity vs discharge current at different ambient temperatures. T1 represents the coldest temperature, T4 represents room temperature, and T6 represents the hottest temperature. Each increasing subscript number (1 through 6) represents battery capacity vs discharge current behavior at a greater ambient temperature.

Now that we're familiar with the graph, let's evaluate each possible answer choice in order.

Notice that except for a very small portion of curves T5 and T6 at small discharge current values, in general, discharge current always increases to the right for all curves as battery capacity decreases in the downward direction.

> True! **(A) In general, an increase in current delivered by a battery will result in a decrease in the total amount of amp hours the battery is able to deliver.**

Compare either of the two capacity vs discharge current curves that are **higher than** room temperature (T5 and T6) and notice that battery capacity **increases** for the same value of discharge current compared to the room temperature capacity vs discharge current curve (T4).

> True! **(B) In general, a battery in an environment with an ambient temperature higher than room temperature will typically result in an increase in the total amount of amp hours the battery is able to deliver.**

Compare any of the three capacity vs discharge current curves that are **less than** room temperature (T1, T2, and T3) and notice that battery capacity **decreases** for the same value of discharge current compared to the room temperature capacity vs discharge current curve (T4).

[29] *NCEES® Reference Handbook (Version 1.1.2) - 3.2.1.7 Relationship of Temperature on Battery Capacity p. 43*

False! **(C) A battery in an environment with an ambient temperature lower than room temperature will result in an increase in the total amount of amp hours the battery is able to deliver.**

Compare the relationship between discharge current and increased temperature by picking any two curves. Notice that for the same capacity, the higher temperature curve will always have a greater discharge current.

True! **(D) A battery will supply more DC current in hotter temperatures at a constant amp hour capacity.**

Compare the relationship between battery capacity (amp hours) and increased (hotter) temperature by picking any two curves. Notice that for the same discharge current, the higher temperature curve will generally always have more battery capacity (amp hours) compared to the lower temperature curve.

True! **(E) A battery will supply more amp hours in hotter temperatures at a constant discharge current.**

Notice that only curves T5 and T6 are able to supply 100% rated capacity (or greater), both of which are hotter than room temperature (T4). At room temperature, battery capacity at greatest is slightly less than 100% rated capacity and continues to decrease as discharge current increases.

True! **(F) A battery can only be expected to supply 100% rated capacity at higher than room temperature.**

Since the problem asks for which of the statements are false, the answer is C.

The answer is: **C.**

(Multiple correct AIT question) - Ch. 5.1 Battery Characteristics and Ratings

Problem #28 Solution

To answer this question correctly, you'll need to reference the *uncontrolled single-phase half-wave rectifier* circuit and the resulting voltage *output graph*[30] in the *NCEES® Reference Handbook*.

An **uncontrolled** rectifier is one that uses diodes to rectify voltage, since diodes cannot be controlled like thyristors, silicon controlled rectifiers (SCRs), insulated-gate bipolar transistors (IGBTs), or other switching electronics.

Let's evaluate each possible answer choice in order.

The formula for peak to peak output ripple voltage[31] is:

$$\Delta V_o = \frac{V_m}{fRC}$$

where:

ΔV_o = Peak to peak ripple voltage [V].
V_m = Maximum (peak) output voltage [V].
f = Frequency [Hz].
R = DC load resistance [Ω].
C = Capacitance of the parallel smoothing capacitor [F].

Notice that as the capacitance (C) of the parallel smoothing capacitor in the unit of farad [F] increases, the peak to peak ripple voltage (ΔV_o) will decrease by the same amount. By definition, this is an inversely proportional relationship.

True! **(A) The peak to peak ripple voltage is inversely proportional to the farad rating of the capacitor.**

We can rewrite the above formula to solve for the peak to peak ripple voltage (ΔV_o) as a percentage of the maximum output voltage (V_m):

$$\frac{\Delta V_o}{V_m} = \frac{1}{fRC} \times 100\%$$

[30] *NCEES® Reference Handbook (Version 1.1.2) - 3.2.2.1 AC-DC Converters p. 44-45*
[31] *NCEES® Reference Handbook (Version 1.1.2) - 3.2.2.1 AC-DC Converters p. 45*

Notice that this value will decrease by the same amount of any increase to the DC load resistance (R), and will similarly increase by the same amount of any decrease to the DC load resistance (R). This is an inversely proportional relationship.

True! **(B) The peak to peak ripple voltage as a percentage of the maximum output voltage will decrease proportionally as the DC load resistance increases.**

The time constant (τ) is the product of resistance (R) and capacitance (C)[32]:

$$\tau = RC$$

Let's substitute this value into the peak to peak output ripple voltage formula:

$$\Delta V_o = \frac{V_m}{fRC}$$

$$\Delta V_o = \frac{V_m}{f\tau}$$

Notice that the time constant (τ) is **inversely proportional** to the peak to peak ripple voltage (ΔV_o), **not** directly proportional.

False! **(C) The time constant of the circuit is directly proportional to the peak to peak ripple voltage.**

We saw earlier that as the capacitance (C) rating of the capacitor increases, the peak to peak ripple voltage (ΔV_o) decreases. Looking at the graph in the Reference Handbook, we can see that the minimum output voltage (V_{min}) is equal to the difference of the maximum peak output voltage (V_m) and the peak to peak ripple voltage (ΔV_o):

$$V_{\min} = V_m - \Delta V_o$$

*Note: This formula is **not** shown in the Handbook and Vmin is **not** labeled on the graph.*

Looking at the graph, we can see that when the output voltage equals the minimum output voltage (V_{min}), the capacitor begins charging until the output voltage reaches the maximum peak output voltage (V_m).

32 *NCEES® Reference Handbook (Version 1.1.2) - 3.1.5 DC Circuits p. 37*

Since the maximum peak output voltage (V_m) remains constant and is equal to the maximum peak output voltage of the AC input voltage, the result is that the minimum output voltage will increase in value and continue to approach the maximum peak output voltage (V_m) as ripple continues to **decrease** as a result of the capacitance rating of the capacitor **increasing**.

> True! **(D) As the capacitance rating of the capacitor increases, the minimum output voltage increases.**

Looking at the circuit in the Handbook, the AC current that passes through the diode (i_D) is supplied by the AC power source only when the capacitor is charging. During discharging cycles, the capacitor supplies all DC current to the DC load (i_R) and the rectifier circuit is open circuited from the capacitor and load due to the diode operating in the reverse bias region while the capacitor is discharging.

> Note: *for a detailed explanation and accompanying illustrative graphs of charging and discharging capacitor cycles in DC rectifier circuits, please see the solution for problem #45 from the* ***Electrical Engineering PE Practice Exam and Technical Study Guide by Zach Stone P.E.***

As the capacitance (C) rating of the capacitor increases, the minimum voltage value increases as we saw previously and, as a result, the capacitor spends more time discharging than it does charging.

Since the AC current that passes through the diode (i_D) is supplied by the AC power source only when the capacitor is charging, there will be a decrease in overall diode current.

> True! **(E) Diode current decreases as the capacitance rating of the capacitor increases.**

Since the problem asks for which of the statements are true, the answer is A, B, D, and E.

The answer is: **A, B, D, and E.**

(Multiple correct AIT question) - Ch. 5.2 Power Supplies and Converters

Problem #29 Solution

Buck-boost converters[33] are able to operate as either a DC-DC step up converter (boost) or as a DC-DC step down converter (buck) depending on the duty ratio (D) setting.

The formula for the output voltage (V_o) of a buck-boost converter is:

$$V_o = -V_s\left[\frac{D}{1-D}\right]$$

where:

V_o = Output DC voltage.
V_s = Input DC Voltage.
D = Duty ratio.

Notice that if D = 0.5, the output voltage (V_o) will be equal to the input voltage (V_s) except opposite in polarity:

$$V_o = -V_s\left[\frac{0.5}{1-0.5}\right]$$

$$V_o = -V_s\left[\frac{0.5}{0.5}\right]$$

$$V_o = -V_s$$

When the duty cycle (D) is less than 0.5:

The buck-boost converter will behave as a **buck (step down) DC to DC converter** and the output voltage (V_o) magnitude (absolute value) will always be **less than** the input voltage (V_s).

If: D < 0.5
Then: $|V_o| < |V_s|$

When the duty cycle (D) is greater than 0.5:

[33] *NCEES® Reference Handbook (Version 1.1.2) - 3.2.2.2 DC-DC Converters p. 48*

The buck-boost converter will behave as a **boost (step up) DC to DC converter** and the output voltage (V_o) magnitude (absolute value) will always be **greater than** the input voltage (V_s).

If: D > 0.5
Then: $|V_o| > |V_s|$

Since the output voltage (V_o) is opposite in polarity (negative) compared to the input voltage (V_s) when the duty cycle (D) is greater than zero or less than one (0 < D < 1), the buck-boost converter is typically only used in applications where opposite voltage polarity will not interfere with the circuit.

The duty cycle (D) can never be equal to 1, as that results in dividing by zero and V_o increasing to infinity:

$$V_o = -V_s\left[\frac{1}{1-1}\right]$$

$$V_o = -V_s\left[\frac{1}{0}\right]$$

D ≠ 1

The duty ratio values given in the answer choices that can be used for a buck-boost converter to operate as a step up DC to DC converter are C (0.6) and D (0.8), since these values are all greater than 0.5 and not equal to one.

The answer is: **C and D.**

(Multiple correct AIT question) - Ch. 5.2 Power Supplies and Converters

Problem #30 Solution

Full-bridge inverters[34] work by rapidly inverting the DC supply voltage (*VDC*) in order to produce an inverted DC to AC output signal.

The four switches operate in pairs.

When switch S1 and S2 are closed and switch S3 and S4 are open, positive DC voltage is applied across the load:

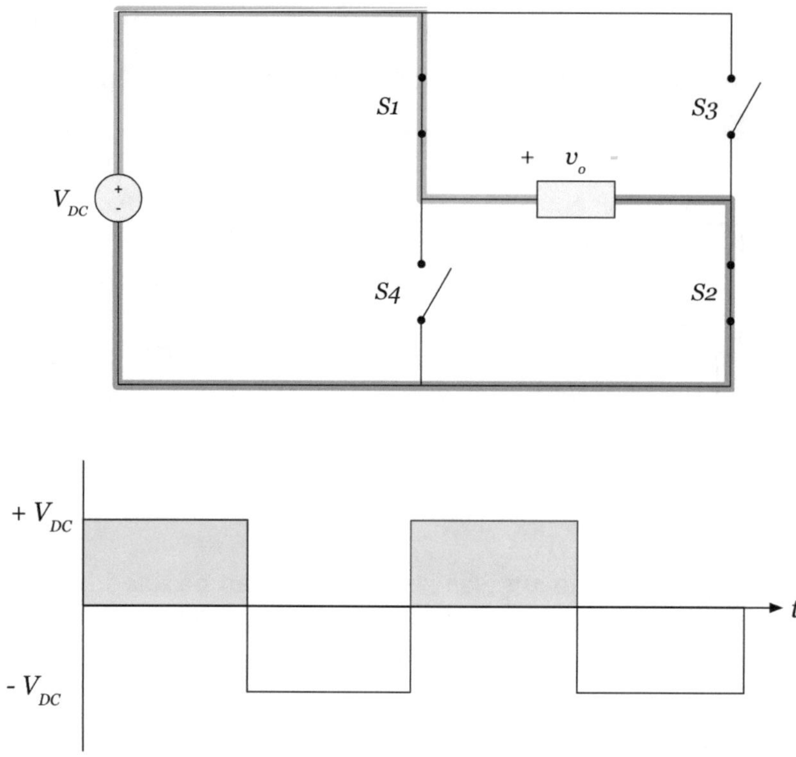

When switch S3 and S4 are closed and switch S1 and S2 are open, negative DC voltage is applied across the load, reversing the voltage polarity:

[34] *NCEES® Reference Handbook (Version 1.1.2) - 3.2.2.3 Inverters p. 48*

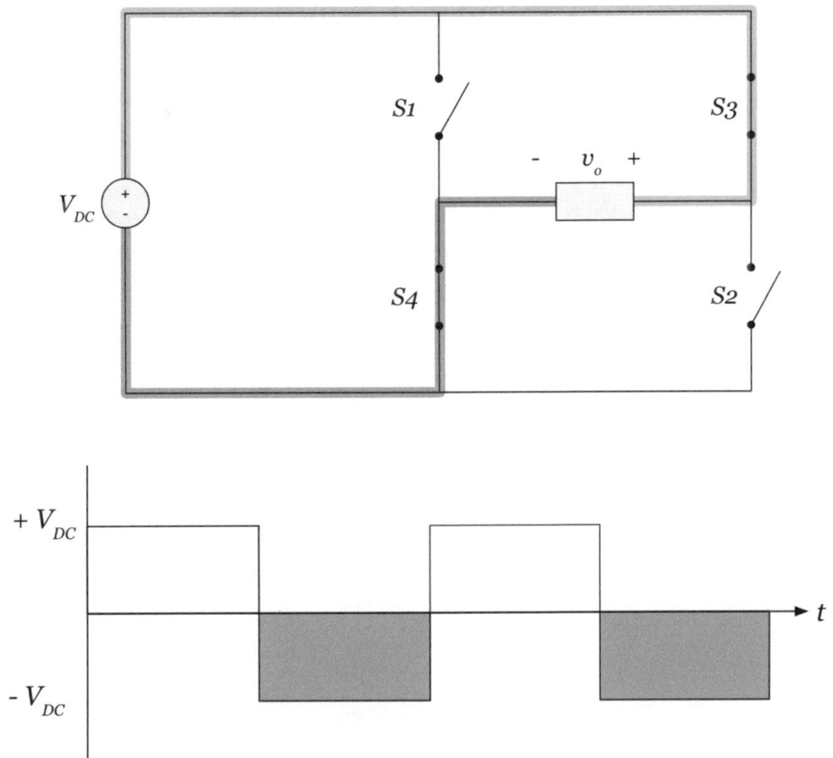

The correct combination of switches in the full-bridge inverter circuit that are closed during the shaded region of the DC voltage graph shown in the problem are switches S3 and S4.

The answer is: **S3 and S4.**

(Point and click AIT question) - Ch. 5.2 Power Supplies and Converters

Problem #31 Solution

For speed control, variable frequency drives (VFDs) are typically designed to keep the applied voltage ratio constant with any changes in frequency. This is known as the **volts per hertz ratio**[35] (V/f).

In order to avoid **magnetic saturation**, the volts per hertz ratio needs to be kept constant such that any changes in motor speed from an increase or decrease in the applied frequency (f) needs to be met with an equal increase or decrease in the applied voltage (V).

Operating a motor without keeping the volts per hertz ratio constant will almost always result in magnetic saturation in the ferromagnetic iron stator core.

The **rotating synchronous magnetic field** speed (n_S)[36] inside a motor stator core is directly proportional to the applied frequency (f):

$$n_s = \frac{120f}{p}$$

In order to safely operate a motor at 50% reduced speed (by decreasing the frequency by 50%), the voltage applied to the motor must also be decreased by the same 50%.

All power devices that operate with a ferromagnetic core (usually iron) such as transformers and rotating machines are susceptible to magnetic saturation.

Magnetic saturation occurs when the applied voltage results in an exponential increase in the current drawn by the machine due to near 100% alignment of the magnetic dipoles in the ferromagnetic core:

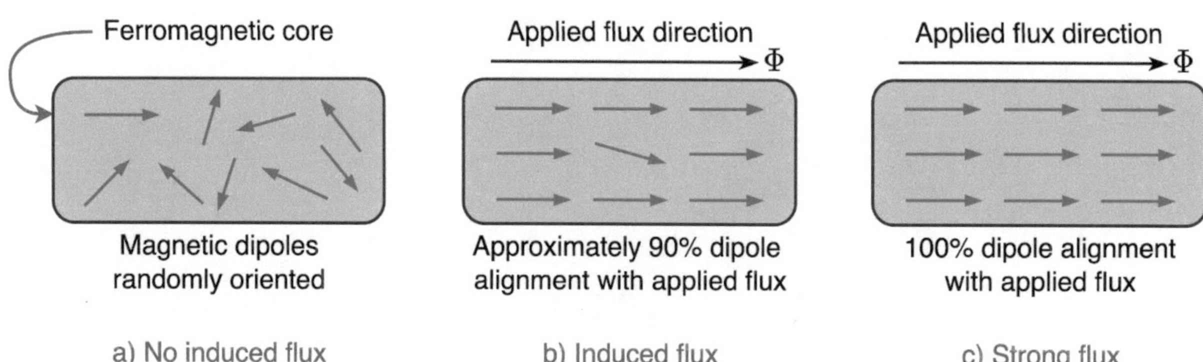

[35] *NCEES® Reference Handbook (Version 1.1.2) - 3.2.3 Variable-Speed Drives p. 48*
[36] *NCEES® Reference Handbook (Version 1.1.2) - 4.1.1.2 Synchronous Speeds p. 50*

Prior to applying voltage (or excitation current), there is no external flux in the core. Without an externally induced flux, the natural dipoles present in ferromagnetic material are randomly aligned (see figure a above).

Once flux is induced, the dipoles begin to orient themselves in the same direction as the externally induced flux. Under typical operating conditions, about 90% of dipoles will align themselves in the same orientation as the induced flux (see figure b above).

If a strong enough flux is induced into the core, close to 100% of the dipoles align with the flux and magnetic saturation occurs (see figure c above).

Let's look at the voltage vs current curve for the voltage applied to the motor by the VFD and the resulting current drawn by the motor to compare:

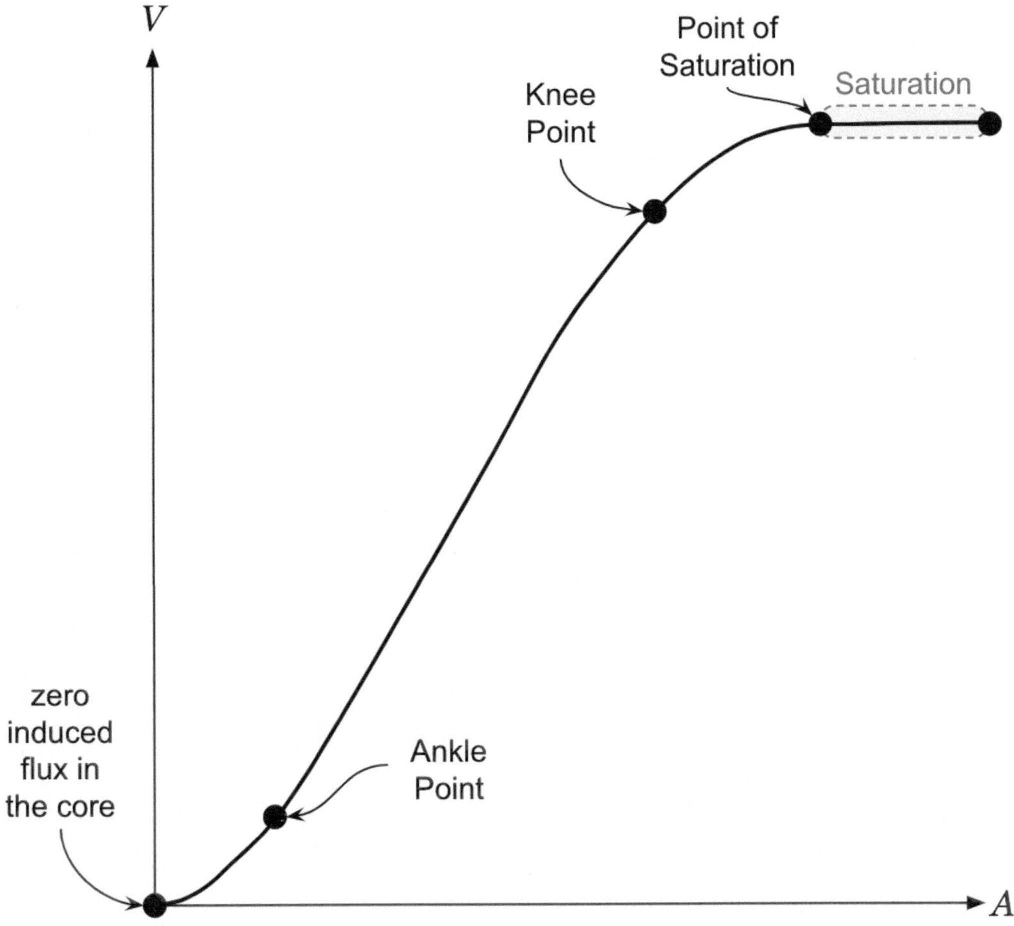

From the **ankle point** to **the knee point**, an increase in applied voltage (or excitation current) results in an approximately linear 1 to 1 increase in the current drawn by the machine.

When the knee point is reached, the relationship is no longer linear and the current drawn by the motor will increase by 50% with just a 10% increase in voltage; this results in reaching the **point of saturation.**

After the point of saturation is reached, the core has reached magnetic saturation.

Most machines operate slightly below the knee point. The knee point is avoided because once it is reached, the current will dramatically increase to the point of saturation with even just a small increase of voltage.

The portion of the above graph that most likely represents the applied voltage (V) and motor current (A) behavior if the VFD applies rated motor voltage is circled in red and labeled "saturation".

The answer is: **see the saturated region of the graph above circled in red.**

(Point and click AIT question) - Ch. 5.4 Variable-speed drives

Problem #32 Solution

There are several different torque formulas in the Reference Handbook to choose from. We will solve for the torque of this motor using two different Handbook torque formulas for extra practice, familiarity, and verification of our work.

The first torque formula[37] we will use is:

$$P(kW) = T(N \cdot m) \cdot n(rpm)/9549$$

where:

P(kW) = Output (mechanical) power of the motor in kilowatts [kW].
T(N•m) = Output (mechanical) torque of the motor in newton meters [N•m].
n = Rotor speed in revolutions per minute [rpm].

Let's rearrange the formula to solve for torque:

$$T(N \cdot m) = \frac{9549 \cdot P(kW)}{n(rpm)}$$

First, we need to solve for the motor output power in kilowatts *P(kW)*. We can do this by converting the horsepower to watts using the following conversion factor[38]:

$$HP\ X\ 745.7 = watt\ [W]$$
$$100\ HP\ X\ 745.7 = 74.57\ kW$$

```
DEG
100*745.7   74570
```

P(kW) = 74.57 kW

Remember, we need this value in kilowatts [kW] for this formula, so be sure to convert from watts to kW.

[37] *NCEES® Reference Handbook (Version 1.1.2) - 4.1.1.1 Power, Torque, and Speed Relationships p. 50*
[38] *NCEES® Reference Handbook (Version 1.1.2) - 1.2 Conversion Factor p. 2*

We can also solve for the motor output power in kilowatts *P(kW)* by calculating the three-phase input power ($P_{in\text{-}3ø}$)[39] using the rated voltage, full load amps, and power factor (*PF*), and then converting to output power using efficiency (η)[40]:

$$P_{in-3ø} = \sqrt{3}|V_L| \cdot |I_L| \cdot PF$$
$$P_{in-3ø} = \sqrt{3}(208V)(241.1A)(0.90)$$

```
DEG
√3*208*241.1*0.9
        78174.24258
```

$$P_{out-3ø} = P_{in-3ø} \cdot \eta$$
$$P_{out-3ø} = 78,174\,W(95.4\%)$$

```
DEG
ans*95.4%
        74578.22743
```

P(kW) = 74.58 kW

Note: the values for P(kW) differ by one hundredth. You may use either value and still answer this question with the same level of rounded accuracy.

Now that we know the motor output power in kilowatts P(kW), we can plug the motor's rated speed, also known as the actual rotor speed (n), from the motor nameplate into the problem to calculate the motor torque using the first torque formula:

$$T(N \cdot m) = \frac{9549 \cdot P(kW)}{n(rpm)}$$

$$T(N \cdot m) = \frac{9549 \cdot 74.58\,kW}{3570\,rpm}$$

```
DEG
9549*ans/1000
     3570
        199.4810907
```

[39] *NCEES® Reference Handbook (Version 1.1.2) - 3.1.1 3-Phase Circuits p. 33*
[40] *NCEES® Reference Handbook (Version 1.1.2) - 2.2.5 Energy Management p. 27*

> ***Careful!*** *To use this formula, the motor output power needs to be in kilowatts [kW], and* ***not*** *watts [W] which is typically more common. This is why we divided the results of the last step in the calculator (ans = 74578.22743) by 1,000.*

The total torque of the motor using the nameplate data rounded to the nearest newton meter is 199 Nm.

The answer is: **199**

Let's solve the same problem using second following torque formula[41] in the Reference Handbook:

$$T_{\text{ind}} = \frac{P_{\text{conv}}}{w_m} \quad \text{and} \quad w_m = (1-s)w_{\text{sync}}$$

where:

P_{conv} = Output mechanical converted power of the motor in watts $[W]$.
s = Motor slip [unitless].
w_{sync} = Synchronous speed of the motor in radians per second $[rad/s]$.

We'll need to calculate slip (s), and synchronous speed (w_{sync}) in units of radians per second $[rad/s]$. We already know the output power of the motor (P_{conv}) from the first solution.

First, let's calculate slip (s)[42]:

> *Note: We aren't given synchronous speed of the motor (n_s) but we can safely assume that it is 3,600 rpm since the actual rated motor speed (n = 3,570 rpm) is always* ***slightly less*** *than the motor's synchronous speed (n_s), and the number of poles (p) in the synchronous speed formula ($n_s = 120f/p$) has to be a whole even number (example: 2, 4, 6, etc).*
>
> *If you're not comfortable making the jump with this assumption, try plugging in different whole even numbers for the number of poles (p) using 60 Hz for the frequency (f) until you arrive at the synchronous speed (n_s) value of 3,600 rpm, a value slightly larger than the motor's rated speed (n), using p = 2.*

---

[41] *NCEES® Reference Handbook (Version 1.1.2) - 4.2.4 Electrical Machine Theory p. 57*
[42] *NCEES® Reference Handbook (Version 1.1.2) - 4.2.1.3 Percent Slip in Induction Machines p. 52*

$$s = \frac{n_s - n}{n_s}$$

$$s = \frac{3600 - 3570}{3600}$$

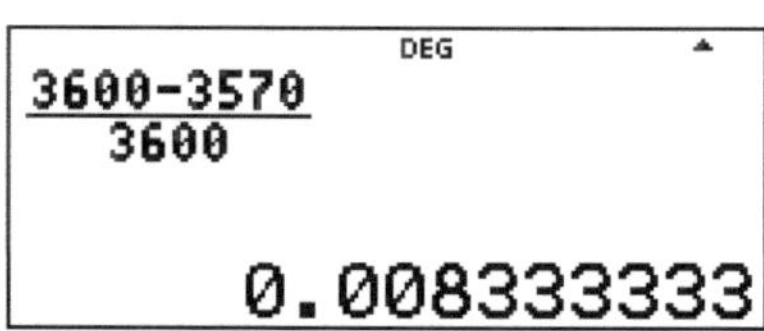

The motor's slip is approximately 0.00833, or 0.833%.

Next, we already know the motor's synchronous speed *(n_s)* in rpm from the previous step, but we'll need to convert it to units of radians per second *[rad/s]* to use it in the torque formula.

We can convert revolutions to radians using the following conversion:

1 revolution = 2π radian

Note: this conversion factor is not included in the Reference Handbook but can be derived from the given radian to degree conversion factor[43] if you are familiar with the relationship that one revolution is equal to 360 degrees and, likewise, half of a revolution is equal to 180 degrees.

We can convert minutes to seconds using the following conversion:

1 min = 60 seconds

Let's convert the synchronous speed of the motor from revolutions per minute [*rpm*] to radians per second [*rad/s*]:

$$w_{\text{sync}} = \frac{\text{revolutions}}{\text{min}} \cdot \frac{2\pi\ \textbf{rad}}{1\ \text{revolution}} \cdot \frac{1\ \text{min}}{60\ \text{sec}}$$

$$w_{\text{sync}} = n_s \cdot \frac{2\pi\ \textbf{rad}}{60\ \text{sec}}$$

$$w_{\text{sync}} = 3,600\ \text{rpm} \cdot \frac{2\pi\ \textbf{rad}}{60\ \text{sec}}$$

[43] *NCEES® Reference Handbook (Version 1.1.2) - 1.2 Conversion Factor p. 2*

```
DEG
3600*2π/60          120π
120π
         376.9911184
```

The synchronous speed of the motor (w_{sync}) in radians per second is 376.99 rad/s.

Finally, let's solve using the torque formula by plugging in the expression for w_m and compare to the value we got using the first torque formula:

$$T_{\text{ind}} = \frac{P_{\text{conv}}}{(1-s)w_{\text{sync}}}$$

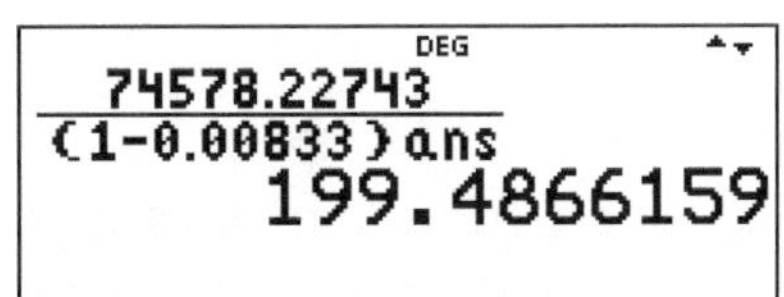

*Careful! Unlike the previous torque formula, the motor output power needs to be in watts [W], and **not** kilowatts [kW].*

The total torque of the motor using the nameplate data rounded to the nearest newton meter is 199 Nm.

The answer is: **199**

(Fill in the blank AIT question) - Ch. 6.1 Electrical Machine Theory

Problem #33 Solution

We can calculate the electrical power delivered by a synchronous machine (P_e), also known as the steady state electrical power, using the following formula[44]:

$$P_e = \frac{3E_0E}{X_d}\sin(\delta) + \frac{3E^2(X_d - X_q)}{2X_dX_q}\sin(2\delta)$$

where:

E_o = The internal or "induced" phase voltage inside the machine.
E = The line-to-neutral rated terminal voltage.
X_d = The direct axis reactance.
X_q = The quadrature axis reactance.
δ = The electrical torque angle.

Unlike cylindrical rotor machines, which make up the vast majority of synchronous machine problems you can expect to see on the PE exam, salient pole rotor machines have two reactance values (direct axis and quadrature) instead of just one.

Before we plug in the values from the problem into our calculator to solve, let's point out a few important details about the variables in this formula.

The internal phase voltage (E_o) **can vary depending on specific operating conditions** of the machine. This is a single-phase line-to-neutral, or "phase", value:

E_o can increase or decrease in magnitude with an increase or decrease in DC excitation.

E_o can increase or decrease in phase angle with an increase or decrease in the amount of mechanical energy supplied to the rotor.

The line-to-neutral rated terminal voltage (E) is a **constant value that does not change** and is a specific machine rating. This is a single-phase line-to-neutral, or "phase", value:

For single-phase machines, E is the voltage rating of the machine.

For three-phase machines, E is the voltage rating of the machine divided by the square root of three.

---

[44] *NCEES® Reference Handbook (Version 1.1.2) - 4.1.4 Electrical Machine Theory p. 51*

The electrical torque angle (δ) is the difference in phase angle between the internal phase voltage (E_o) and the line-to-neutral rated terminal voltage (E):

$$\delta = \theta_{E_0} - \theta_E$$

This is the amount of electrical degrees that the internal phase voltage (E_o) leads or lags the rated line-to-neutral terminal voltage (E) by.

Now we're ready to solve, so let's plug in the values from the problem:

$$P_e = \frac{3E_0E}{X_d}\sin(\delta) + \frac{3E^2(X_d - X_q)}{2X_dX_q}\sin(2\delta)$$

$$P_e = \frac{3(12\ kV)\left(\frac{13.8\ kV}{\sqrt{3}}\right)}{5.12\ \Omega}\sin(15°) + \frac{3\left(\frac{13.8\ kV}{\sqrt{3}}\right)^2(5.12\ \Omega - 3.39\ \Omega)}{2(5.12\ \Omega)(3.39\ \Omega)}\sin(2 \times 15°)$$

You may have to perform this as the sum of two different formulas in your calculator in case you reach the limit of the number of characters you are able to enter:

```
DEG
3(12E3)(13.8E3/√3)/5.12 sin(15
14499306.46
```

```
DEG
ans+ 3(13.8E3/√3)^2(5.12-3.39)/2(5.12)(3.39) sin(2*15
```

```
DEG
19244726.13
```

The electrical power delivered by the machine is 19,244,726 watts. Rounded to the nearest whole unit shown in the problem is 19 megawatts.

The answer is: **19**

(Fill in the blank AIT question) - Ch. 6.1 Electrical Machine Theory

Problem #34 Solution

The variables shown in the completed induction machine equivalent circuit[45] in the Reference Handbook can be used to answer this problem along with the names of each motor circuit element variable[46] and the induction motor power loss formulas and descriptions[47]:

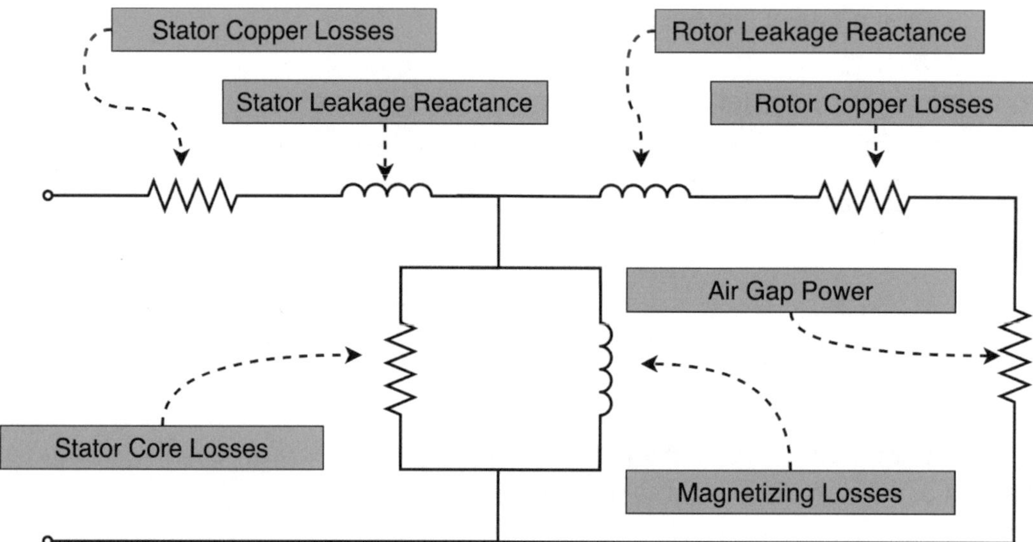

Both the stator leakage reactance (X_1) and the rotor leakage reactance referred to the stator (X_2) contribute to the overall reactive power losses of the induction motor.

The answer is: **see the completed diagram above with label names shown above.**

(Drag and drop AIT question) - Ch. 6.2 Equivalent Circuits and Characteristics

[45] *NCEES® Reference Handbook (Version 1.1.2) - 4.2.2 Equivalent Circuits and Characteristics p. 54*
[46] *NCEES® Reference Handbook (Version 1.1.2) - 4.2 Induction Machines p. 52*
[47] *NCEES® Reference Handbook (Version 1.1.2) - 4.2.4 Electrical Machine Theory p. 57*

Problem #35 Solution

First, let's fill in the induction motor equivalent circuit parameters using the values given in the problem.

Since we will be finding the Thévenin-equivalent resistance (R_{Th}) of just the stator, we'll need to use the first induction motor equivalent circuit in the Reference Handbook[48] that separates the stator impedance (R_1 and jX_1) from the rotor impedance (R_2 and jX_2) with the parallel magnetizing branch impedance (R_C and jX_M) separating the stator and rotor:

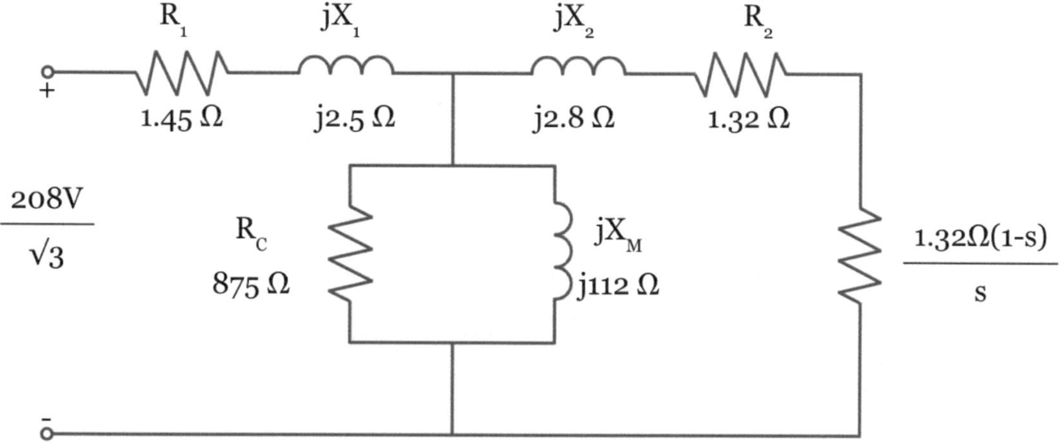

The Thévenin-equivalent impedance (Z_{Th}) of the stator to the power supply can be calculated by shorting the applied voltage, then calculating the equivalent impedance at the terminals of the circuit where the stator meets the rotor:

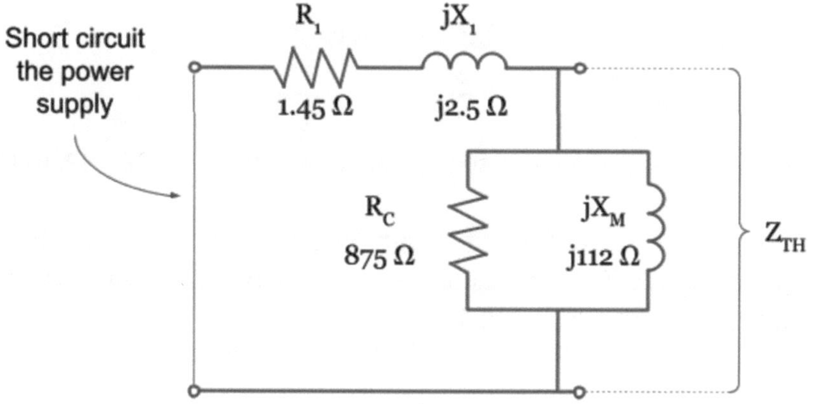

The Thévenin-equivalent impedance (Z_{Th}) of the stator is:

[48] *NCEES® Reference Handbook (Version 1.1.2) - 4.2.2 Equivalent Circuits and Characteristics p. 54*

$$\hat{Z}_{Th} = (R_1 + jX_1)//[(R_C//jX_M)]$$

$$\hat{Z}_{Th} = (1.45 + j2.5\,\Omega)//[(875\,\Omega//j112\,\Omega)]$$

The easiest way to do this in your calculator is in parts by storing values to variables. Let's store the series stator winding impedance (1.45 + j2.5 Ω) in the calculator to variable "x":

```
DEG
1.45+i2.5→x
        1.45+2.5i
```

Next, let's calculate the equivalent parallel branch impedance and store it in the calculator to variable "y":

```
DEG
875*i112
-------- →y
875+i112
14.10490523+110▸
```

Finally, let's use the "x" and "y" variables to calculate the Thévenin-equivalent impedance (Z_{Th}) of the stator:

```
DEG
x*y
---
x+y
1.391854395+2.4▸
```

The Thévenin-equivalent impedance (Z_{Th}) is approximately 1.39 + j2.46 Ω.

Since the problem asks for the Thévenin-equivalent resistance (R_{Th}), we can easily determine it from the real component of the Thévenin-equivalent impedance (Z_{Th}) which is approximately 1.39 Ω, or rounded to the nearest ohm just 1Ω.

The answer is: **1.**

(Fill in the blank AIT question) - Ch. 6.2 Equivalent Circuits and Characteristics

Problem #36 Solution

The correct NEMA design letter for each application described in the problem is:

NEMA E	High efficiency applications.
NEMA C	High starting torque applications with standard slip ranges.
NEMA D	High starting torque applications with a need for large values of slip.
NEMA B	Constant speed applications with reduced starting current.
NEMA F	Low starting current applications.
NEMA A	Constant speed applications with high starting current.

This is a difficult problem to answer because the *NCEES® Reference Handbook* does not include information about the different types of NEMA motor applications.

While the majority of what you will need to pass the PE exam will be included in the *NCEES® Reference Handbook*, you can expect that there will still be a small number of problems that cannot be answered using the Reference Handbook and must be solved either from field experience in the industry or from information learned while studying for the PE exam.

NEMA design letters for motors are the most fundamental aspect of selecting the correct motor for specific applications since each NEMA design letter will have a different torque vs speed curve. Design letter A and B motors tend to be the most commonly used motors for standard applications, while design letter C and D are more suited for applications requiring higher starting torque.

Below is a brief description of the NEMA design letters from the *Electrical PE Review, Inc On-demand Review Course* to help familiarize you with the different types.

> **Design letter A motors** - Standard squirrel cage induction motor designed for constant speed. Starting torque is 1.5 to 1.75 times the rated full load torque. Design letter A has the best speed regulation of all NEMA design letter motors, but has a drawback of the highest starting current ranging from 5 to 7 times the normal rated full load current, making it less advantageous for across the line starting applications.

Design letter B motors - Standard starting torque motor. Design letter B has a similar slip torque curve compared to design letter A but has an increased starting and run reactance resulting in reduced starting torque and also reduced starting current. It is considered a general purpose motor and is typically used over design letter A except in smaller sizes. Design letter B motors have a starting current ranging from 4.5 to 5 times the normal full load rated amps and typically utilize reduced voltage starting.

Design letter C motors - High starting torque motor. A double cage induction rotor with a starting torque of 2 to 2.5 times rated torque and a starting current of 3.5 to 5 times rated current, which is much lower compared to both design letter A and design letter B motors. Design letter C is a good fit for sudden large loads with low starting inertia since the high starting torque results in rapid acceleration and heavy high inertia loads that would cause thermal issues in the stator windings, especially with frequent starts. This design letter motor is not suited for constant speed regulation as it will continue to develop increased torque from the resulting increased slip starting from standstill to maximum torque.

Design letter D motors - High slip motor. This is both a high torque and high resistance motor with a higher starting ratio of rotor resistance to reactance than the previous design letter motors. This motor will reach 3 times the rated torque with a starting current 3 to 8 times the rated full load amps. This is the best fit for heavy starting duty with high sudden inertia, such as loaded conveyor belts, as long as there are no frequent starts which would also cause thermal issues. Design letter D has the worst speed regulation of all 6 design letter motors described here.

Design letter E motors - The highest efficiency motor with the lowest losses of all of the design letters described here. This motor has a higher inrush current compared to design letter A and design letter B.

Design letter F motors - A double cage low torque induction motor suited for low starting current applications. This design letter has the highest resistance of all NEMA design letter motors which results in increased starting and running impedance with reduced starting and running current. Design letter F motors are designed as replacements for design letter B motors as the starting torque is 1.25 times the rated torque while sustaining a low starting current of 2 to 4 times the rated full load amps. The benefit of design letter F motors over design letter B motors is not needing reduced voltage starting while allowing direct across-the-line voltage starting due to the low starting current. The drawbacks compared to design letter B are diminished speed regulation, low overload capability, and poor running efficiency.

The answer is: **see the completed drag and drop diagram above.**

(Drag and drop AIT question) - Ch. 6.1 Generator and Motor Applications

Problem #37 Solution

The correct label for each point on the induction machine torque vs speed curve is shown below:

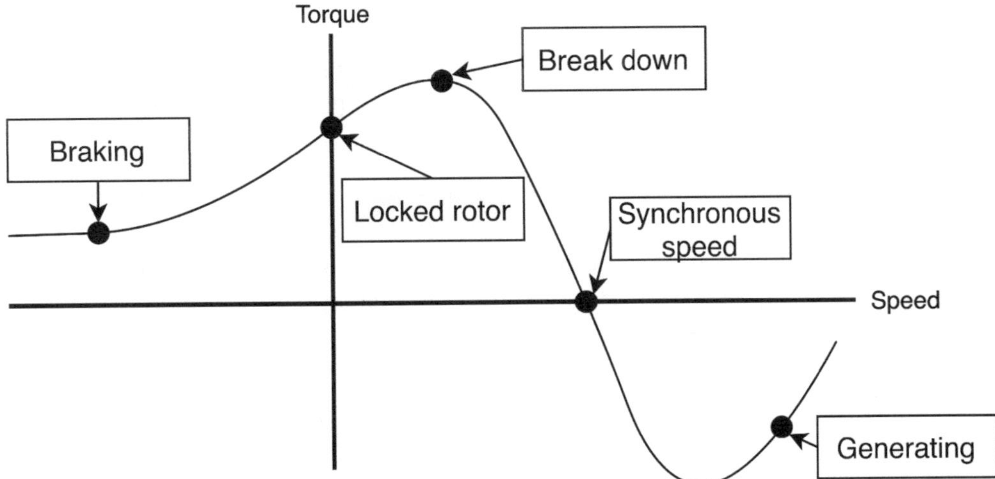

The *NCEES® Reference Handbook* has the motoring portion of the torque vs speed curve for a particular machine[49] but it does not include the braking portion of the curve that occurs when rotor speed is less than zero ($n < o$), nor the generating portion of the curve that occurs when rotor speed is greater than synchronous speed ($n > n_s$).

The torque vs speed curve will always be slightly different depending on the design of the machine but will generally follow the same trend.

Below is a brief description of each of the points in the above diagram from left to right:

Braking: When the rotor speed is less than zero ($n < o$), the direction of the rotating magnetic field in the stator windings changes direction. This causes the machine to slow down and come to a stop much faster compared to de-energizing the power supply and allowing the machine to coast to a stop. When braking, the power supply must be de-energized when the rotor comes to a stop, otherwise the rotor will start to motor in the opposite direction.

Locked rotor: Occurs when the rotor speed is equal to zero ($n = o$) at the intersection of the torque vs speed curve and the vertical torque axis. Locked rotor occurs when the rotor stops suddenly during operating conditions when the power supply is still energized. It also occurs momentarily when the machine is first started from rest.

Break down: The point on the torque vs speed curve when torque is at the maximum.

[49] *NCEES® Reference Handbook (Version 1.1.2) - 4.2.4 Electrical Machine Theory p. 56*

Synchronous speed: When the rotor speed is equal to the synchronous speed of the rotating magnetic field in the stator ($n = n_s$). This occurs at the intersection of the torque vs speed curve and the horizontal speed axis. When this occurs, torque is equal to zero. This could represent either an unloaded motor or an unloaded generator.

Generating: When the rotor speed is greater than synchronous speed ($n > n_s$), the slip of the machine changes from positive to negative as the rotor spins faster than the synchronous speed of the rotating magnetic field in the stator. When this occurs, the machine will supply real power (P, watts) to the connected power system.

Not included in the possible answer choices is **motoring**. This occurs when the rotor speed is greater than zero ($n > o$) starting from locked rotor conditions up until just before the rotor speed is equal to synchronous speed ($n = n_s$).

The answer is: **see the completed drag and drop diagram above.**

(Drag and drop AIT question) - Ch. 6.1 Generator and Motor Applications

Problem #38 Solution

Reduced voltage starting methods are used to limit the starting current and the starting torque of induction motors. The typical applications of reduced voltage and current-limited starting are for mechanical loads that require gradual starting with a slower ramp up of torque compared to full voltage starting.

A classic example of a reduced voltage starting application is a heavily loaded conveyor belt started from rest that could either break with full voltage starting or would draw such large starting current that the motor may overload and trip protection devices.

Other applications for reduced voltage starting are when the starting of a large motor may cause a voltage dip significant enough to damage other equipment powered by the same electrical system.

Current drawn by a motor is **directly proportional** to the voltage applied to it according to Ohm's law ($I=V/Z$). This means that a reduction in applied voltage (V) will result in an equal reduction in current (I) drawn by the motor.

However, unlike the linear relationship with current, the power drawn by the motor will be reduced by a factor of the square of the applied voltage ($S = V^2/Z$). Since torque is **directly proportional** to power ($P(kW) = T(N\cdot m) \cdot n(rpm)/9549$), this means that torque will also be reduced by a factor of the square of the applied voltage.

Let's evaluate each of the possible answer choices to determine which of the motor starting methods may be used to limit the starting current of an induction motor:

(A) Across the line starting

Across the line starting[50] is the most common starting method for an induction motor. A heavy duty motor contactor, often called a motor starter, is energized by closing a physical switch. When the contactor is energized, normally open contacts connecting the motor terminals to the power supply change from open to closed and apply full line voltage across the motor terminals.

The result of applying full line voltage to an induction motor is a large momentary locked rotor inrush current that is many times greater than the rated full load current of the motor at rated speed.

[50] *NCEES® Reference Handbook (Version 1.1.2) - 4.2.3 Motor Starting p. 55*

False! Across the line starting may not be used to limit the starting current of an induction motor.

(B) Reduced voltage auto-transformer starting

Reduced voltage auto-transformer starting[51] steps down the applied voltage during the starting sequence of an induction motor to limit the starting current.

Shortly after the initial starting when the motor is close to rated speed, the auto-transformer is bypassed out of the circuit in order to supply full voltage to the motor while it is running so that the motor can develop rated torque.

True! Reduced voltage auto-transformer starting may be used to limit the starting current of an induction motor.

(C) Reduced voltage resistance starting

Reduced voltage resistance starting[52] uses resistors in series with the power supply and motor in order to reduce the voltage during the starting sequence by creating a temporary voltage drop across the resistor and thereby lowering the amount of voltage at the motor terminals.

With a reduced amount of voltage applied to the motor during starting, the motor will draw less current. Shortly after the initial starting when the motor is close to rated speed, the series resistor is bypassed out of the circuit in order to supply full voltage to the motor while it is running so that the motor can develop rated torque.

True! Reduced voltage resistance starting may be used to limit the starting current of an induction motor.

(D) Wye start delta run

Wye start delta run[53] uses contacts during the motor starting sequence to connect the stator windings in a wye connection to the power supply. Since the applied voltage to the motor during the wye start sequence is line to neutral, the line voltage is reduced by a factor of $\sqrt{3}$.

With a reduced amount of voltage applied to the motor during starting, the motor will draw less current. Shortly after the initial starting, when the motor is close to rated speed, a separate set of contacts is used to change the motor stator connections from

[51] *NCEES® Reference Handbook (Version 1.1.2) - 4.2.3 Motor Starting p. 56*
[52] *NCEES® Reference Handbook (Version 1.1.2) - 4.2.3 Motor Starting p. 55*
[53] *NCEES® Reference Handbook (Version 1.1.2) - 4.2.3 Motor Starting p. 56*

wye to delta in order to apply full line voltage to the motor so that the motor can develop rated torque.

True! Wye start delta run starting may be used to limit the starting current of an induction motor.

(E) Soft start speed drive

A soft starter is very similar to a variable-frequency drive (VFD)[54] except it is only used during the starting sequence. Power electronics inside the soft starter slowly increase the applied voltage of the motor during the starting sequence until reaching full voltage so that the motor may develop rated torque.

The current drawn by the motor will start out low as a decreased amount of voltage is applied to the motor, and will continue to increase as the voltage applied to the motor increases.

True! Soft start speed drives may be used to limit the starting current of an induction motor.

The answer is: **B, C, D, and E.**

(Multiple correct AIT question) - Ch. 6.3 Motor Starting

[54] *NCEES® Reference Handbook (Version 1.1.2) - 3.2.3 Variable-Speed Drives p. 48*

Problem #39 Solution

First, let's draw the parallel transformer circuit filling in the values from the table in the problem:

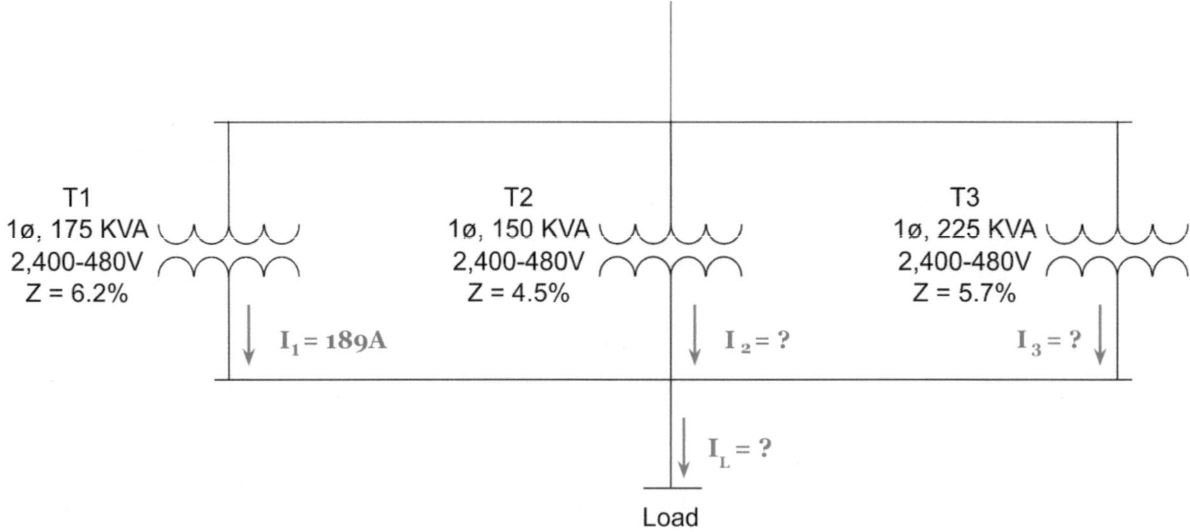

Since the current supplied by transformer T1 (I_1 = 189 A) is given in the problem, we can use this value with the parallel transformer current ratio formula[55] to solve for the current supplied by the remaining two transformers.

First, let's solve for the current supplied by transformer T2 (I_2):

$$\frac{I_1}{I_2} = \frac{(\%Z_{T2})S_{T1}}{(\%Z_{T1})S_{T2}}$$

$$I_2 = I_1 \frac{(\%Z_{T1})S_{T2}}{(\%Z_{T2})S_{T1}}$$

$$I_2 = 189\ A\ \frac{(6.2\%)150\ kVA}{(4.5\%)175\ kVA}$$

[55] *NCEES® Reference Handbook (Version 1.1.2) - 4.3.1.6 Single-Phase Transformers in Parallel p. 61-62*

```
DEG
189*(6.2%*150)/(4.5%*175)
223.2
```

The current supplied by transformer T2 (I_2) is 223.2 A.

Next, let's solve for the current supplied by transformer T3 (I_3):

$$\frac{I_1}{I_3} = \frac{(\%Z_{T3})S_{T1}}{(\%Z_{T1})S_{T3}}$$

$$I_3 = I_1 \frac{(\%Z_{T1})S_{T3}}{(\%Z_{T3})S_{T1}}$$

$$I_3 = 189\ A \frac{(6.2\%)225\ kVA}{(5.7\%)175\ kVA}$$

```
DEG
189*(6.2%*225)/(5.7%*175)
264.3157895
```

The current supplied by transformer T3 (I_3) is approximately 264.3 A.

Note: we are using I_3, $\%Z_{T3}$, and S_{T3} in place of I_2, $\%Z_{T2}$, and S_{T2} in the parallel transformer current ratio formula when solving for the current supplied by the third transformer T3 (I_3). As long as we are consistent with assigning the values for transformer T1 to I_1, $\%Z_{T1}$, and S_{T1}, then we may use the variables I_2, $\%Z_{T2}$, and S_{T2} to solve for the values of any remaining transformers connected in parallel to the same circuit, no matter how many additional parallel transformers there are.

We're now ready to use these values to calculate the total load current (I_L) using Kirchhoff's current law (KCL).

We can assume that the current supplied by each transformer are all in phase with equal phase angles since the problem did not specify the transformer connection type and since parallel transformers typically MUST have the same connection type to avoid secondary currents with different phase angles.

Because of this, we can add the magnitudes of each current in the following KCL formula which typically uses complex numbers since the assumed phase angle of each current is assumed to be the same:

$$I_L = I_1 + I_2 + I_3$$

$$I_L = 189\,A + 223.2\,A + 264.3\,A$$

Since this is a fill in the blank AIT question, be sure to recall all previous values in your calculator to include all decimal places for final rounding accuracy:

```
189+223.2+264.3157895
                676.5157895
```

The total load current (I_L) rounded to the nearest amp is 677A.

Here is the completed diagram with all current values:

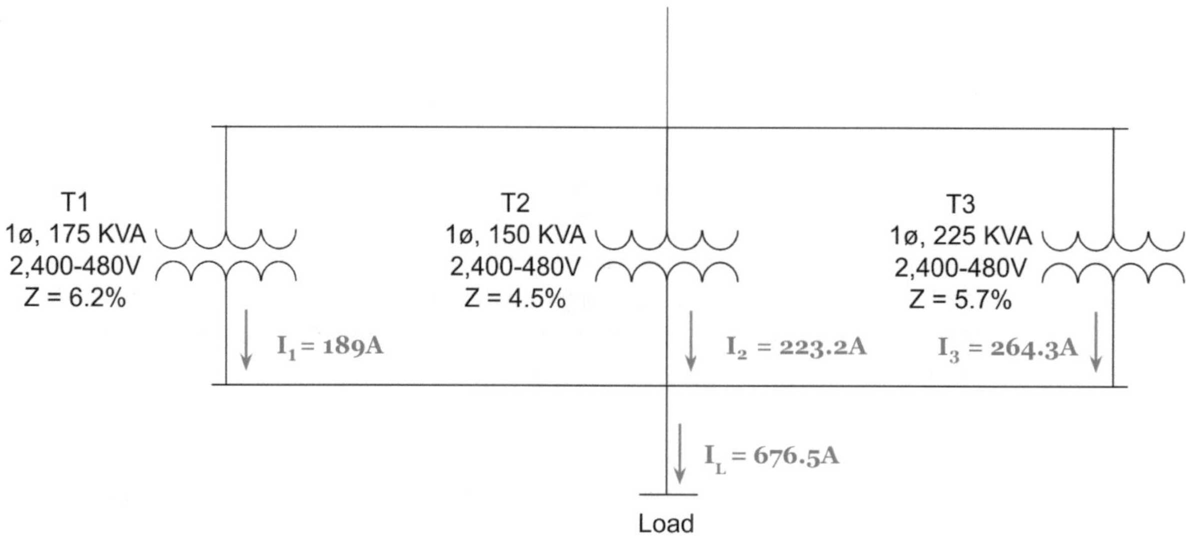

The answer is: **677.**

(Fill in the blank AIT question) - Ch. 7.1 Transformers

Problem #40 Solution

Let's evaluate each of the possible answer choices to determine which ones are true.

(A) The apparent power output of a transformer operating at maximum efficiency can be determined by multiplying the apparent power rating of the transformer by the square root of the ratio of the magnetizing core losses to the full load copper losses.

The formula for transformer percent load at maximum efficiency is[56]:

$$\%load = \sqrt{\frac{P_{core}}{P_{cu}}} \times 100$$

P_{core} is the variable for the magnetizing power losses consumed in the parallel core resistance (R_C) in the transformer equivalent circuit.

P_{cu} is the variable for the copper losses of the transformer at full load consumed in both the primary (R_1) and secondary (R_2) series winding resistances in the transformer equivalent circuit.

True! This formula calculates the percentage of full load (also known as percent load) that results in the transformer operating at maximum (highest) efficiency.

(B) Maximum efficiency of a transformer occurs when the transformer load is equal to the transformer rated power.

We determined in the previous choice that percent load for maximum efficiency can be determined by the following formula:

$$\%load = \sqrt{\frac{P_{core}}{P_{cu}}} \times 100$$

False! Since full load copper losses (P_{cu}) are always much larger than no load core losses (P_{core}), this condition cannot exist at 100% load.

(C) Maximum efficiency of a transformer occurs when the transformer load results in an equal amount of I^2R losses in the transformer compared to the no load losses.

[56] *NCEES® Reference Handbook (Version 1.1.2) - 4.3.1.5 Condition for Maximum Efficiency p. 60-61*

The condition for transformer maximum efficiency[57] is when the transformer is loaded up until the point that the copper losses, which vary with load, are exactly equal to the no load core losses, which are constant:

$$P_{cu} = P_{core}$$

I^2R losses is another term for copper losses (P_{cu}). However, unlike the previous formula, the variable P_{cu} in this relationship does **not** represent the value of the copper losses at full 100% load. Instead, used in this context, P_{cu} will have the same value as the constant no load core losses (P_{core}) only when the transformer is operating exactly at maximum efficiency.

As the percent load of the transformer continues to increase past the point that the load-dependent copper losses are greater than the constant core losses, transformer efficiency will begin to decrease.

True! When a transformer is loaded up to the point that the copper losses (P_{cu}) equal the no load core losses (P_{core}), the transformer is operating at maximum (highest) efficiency.

(D) Transformer copper losses are directly proportional to changes in percent load.

Transformer I^2R copper losses can be found using the general power formula relating real power to current and resistance:

$$P = I^2R$$

You can use either the primary current and the total series winding resistance referred to primary, or the secondary current and the total series winding resistance referred to secondary.

Notice that this is **not** a 1:1 directly proportional relationship.

False! Copper losses are not directly proportional to changes in percent load, they are directly proportional to the **square** of changes in percent load.

(E) As long as the percent load is constant, changes to the load power factor will not have an effect on transformer efficiency.

We can evaluate this using the transformer efficiency (η) formula[58]:

---

[57] *NCEES® Reference Handbook (Version 1.1.2) - 4.3.1.5 Condition for Maximum Efficiency p. 60-61*
[58] *NCEES® Reference Handbook (Version 1.1.2) - 4.3.1.4 Transformer's Efficiency p. 60*

$$\eta = \frac{P_{out}}{P_{out} + P_{cu} + P_c} \times 100\%$$

The output power (P_{out}) of a transformer in the formula above can be calculated using the applied voltage (V), load current (I), and power factor (PF):

$$P_{out} = VI \times PF$$

Substituting this into the efficiency (η) formula:

$$\eta = \frac{VI \times PF}{VI \times PF + P_{cu} + P_c} \times 100\%$$

Notice in the formula above that even when the load current (I) is held constant, changes to power factor (PF) will change the value for transformer efficiency (η).

False! Even with a constant load, changes in power factor will result in different values for transformer output power (P_{out}) which will result in changes to transformer efficiency (η).

The answer is: **A and C.**

(Multiple correct AIT question) - Ch. 7.1 Transformers

Problem #41 Solution

The more the natural cooling processes of the transformer are assisted through devices such as coolant pumps and external air fans, the greater the output power the transformer will be able to supply without overheating.

Although cooling methods of liquid-filled or dry-type transformers are not included in the Reference Handbook, this is a fundamental application of transformers that is fair to be tested on during the PE exam.

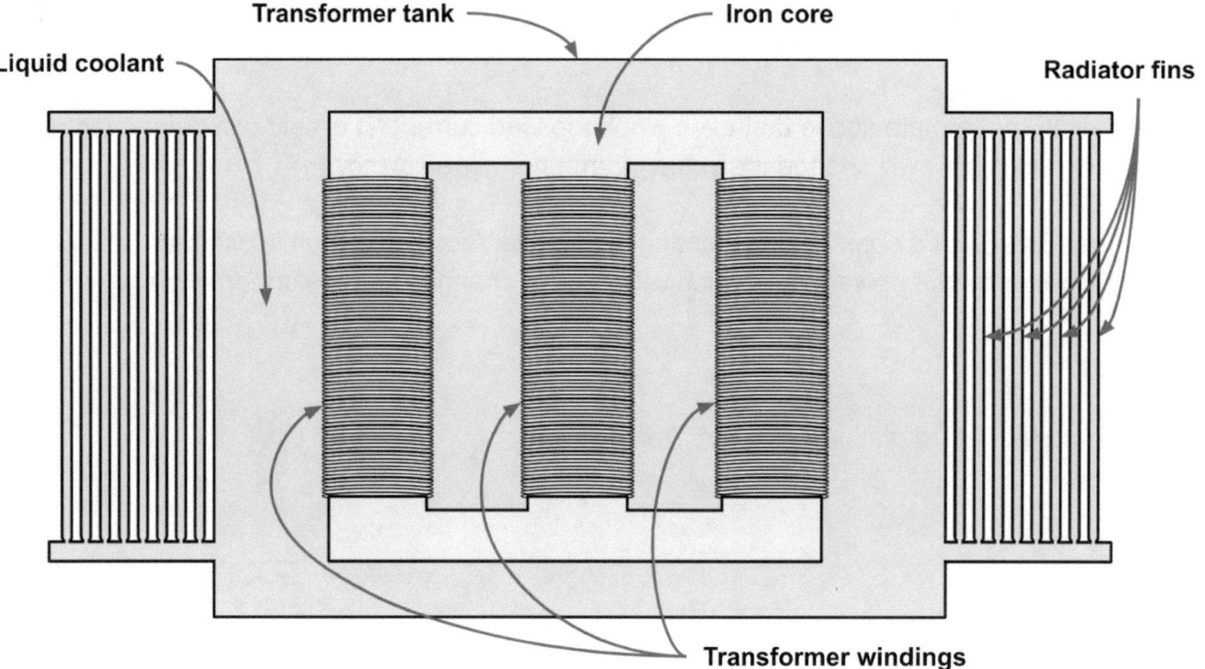

The transformer cooling methods given in this problem ranked from the highest power output capacity to the lowest power output capacity are:

ODAF	4. Highest power output capacity
OFAF	3. Second highest power output capacity
ONAF	2. Second lowest power output capacity
ONAN	1. Lowest power output capacity

Transformers built after the year 2000 use a four letter acronym for each cooling method:

First letter - Internal cooling medium. This is the liquid medium that the transformer is submerged in inside the transformer tank.

O - Liquid coolant with a flashpoint less than or equal to 300°C (oil).

K - Liquid coolant with a flashpoint greater than 300°C (synthetic coolant).

L - Liquid coolant with no measurable flashpoint (synthetic coolant).

Second letter - Internal cooling mechanism. This is the mechanism of how the internal cooling medium circulates inside the transformer tank.

N - Natural liquid coolant convection through cooling equipment and windings.

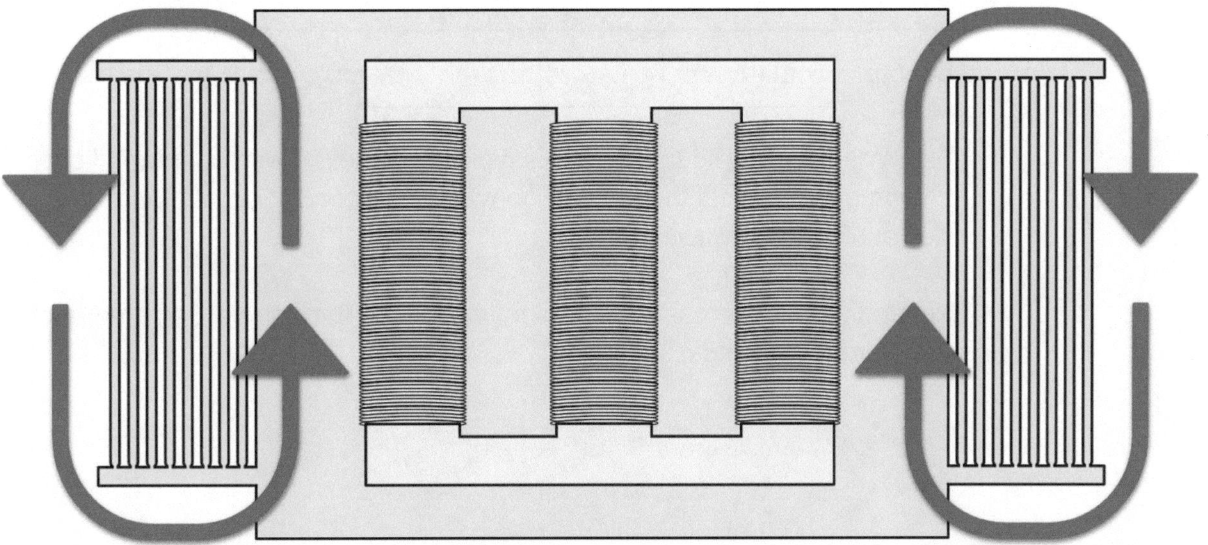

Natural coolant thermal convection in radiator fins due to heat

Natural coolant convection is a result of thermal convection. Hotter oil naturally rises and circulates inside the transformer tank radiator fins.

F - Forced internal liquid coolant circulation through cooling equipment and natural convection in windings.

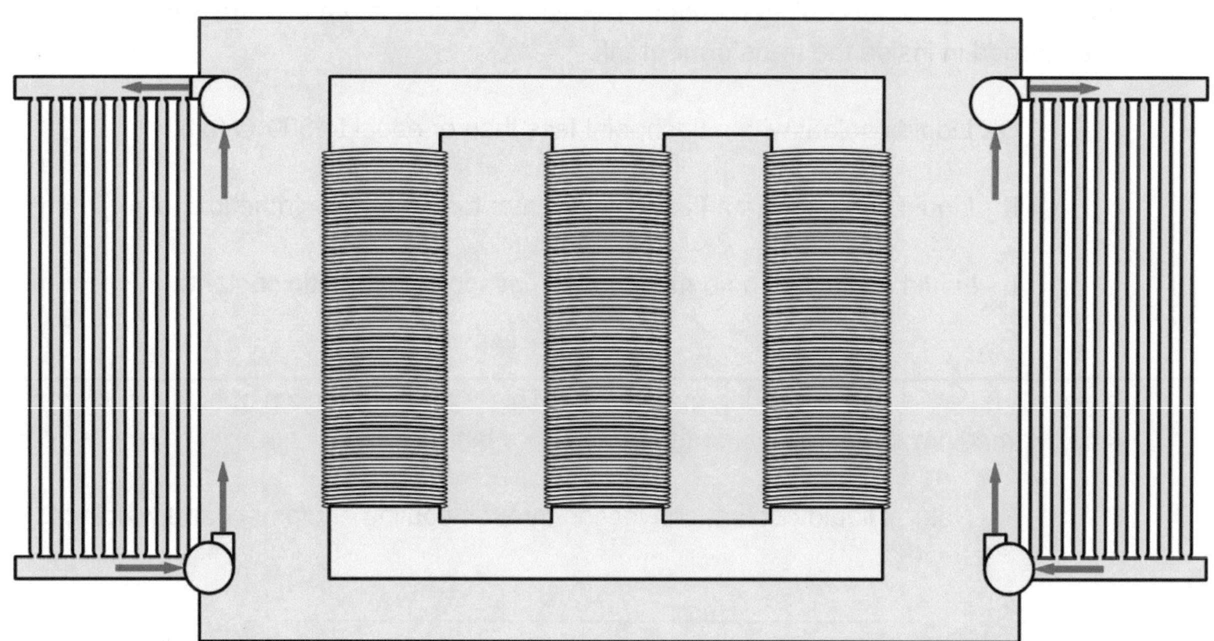

Forced coolant circulation in radiator fins with internal coolant pumps

Forced internal liquid coolant is typically achieved with an internal coolant pump that assists the thermal convection of coolant circulation in the transformer tank radiator fins.

D - Directed forced internal liquid coolant circulation through cooling equipment and flow in main windings.

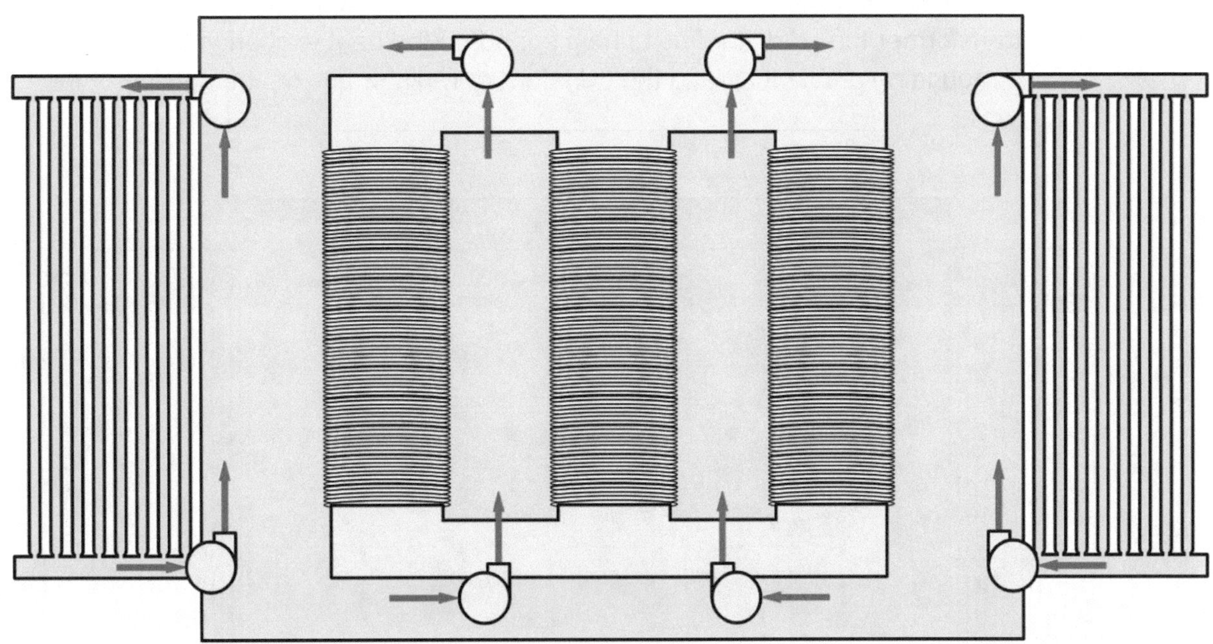

Coolant circulation directed in windings and forced in radiator fins with internal coolant pumps

Directed forced liquid coolant is typically achieved with an additional internal coolant pump that directs the flow of coolant through the transformer windings in addition to a pump that assists the thermal convection of coolant circulation in the transformer tank radiator fins.

Third letter - External cooling medium. This is the medium outside the transformer tank that exchanges heat with the internal cooling medium.

A - Air.

W - Water.

Fourth letter - External cooling mechanism. This is the mechanism of how the external cooling medium exchanges heat with the internal cooling medium.

N - Natural convection.

Natural convection is unassisted.

F - Forced cooling.

Forced cooling with air (A) is typically achieved by external fans mounted on the transformer tank radiator fins to help speed up the heat exchange between the surrounding ambient air and the transformer radiator fins on the outside of the tank.

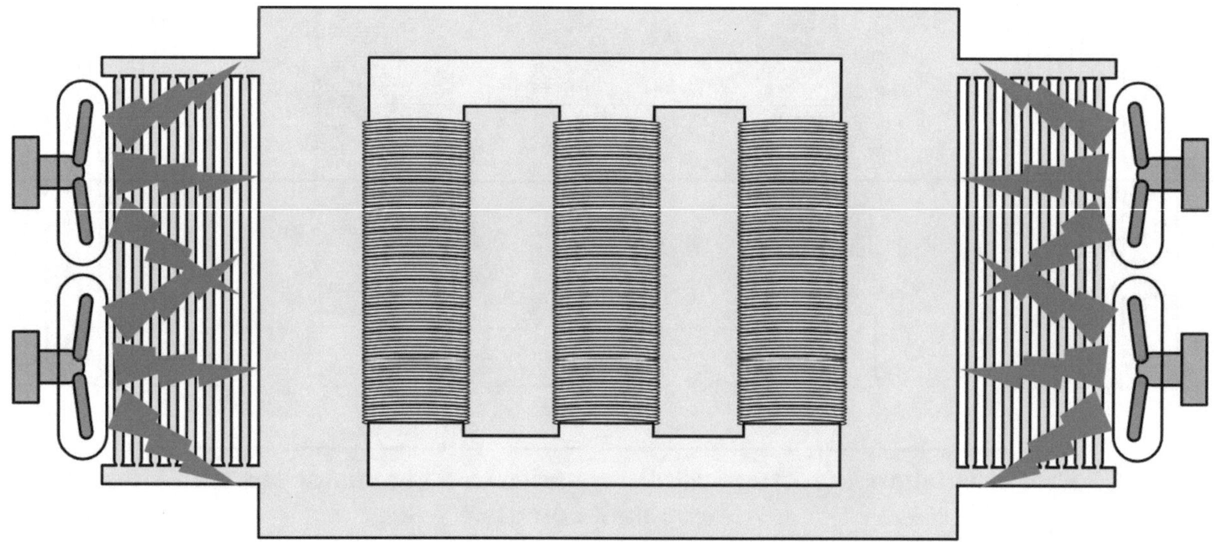

Forced air heat exchange with coolant in radiator fins with external fans

Forced cooling with water (W) (not shown) is typically achieved with additional pumps to help speed up the heat exchange between an external cold water system and the internal liquid coolant.

The cooling methods given in this problem:

ODAF - **O**il **D**irected **A**ir **F**orced

A transformer with a liquid coolant with a flashpoint less than or equal to 300°C (oil) that is being directed through the transformer windings and the radiator fins, with mounted external fans to force the outside ambient air through the radiator fins.

OFAF - **O**il **F**orced **A**ir **F**orced

A transformer with a liquid coolant with a flashpoint less than or equal to 300°C (oil) that is forced through the radiator fins, with mounted external fans to force the outside ambient air through the radiator fins.

ONAF - **O**il **N**atural **A**ir **F**orced

A transformer with a liquid coolant with a flashpoint less than or equal to 300°C (oil) that is naturally flowing through the radiator fins from thermal convection, with mounted external fans to force the outside ambient air through the radiator fins.

ONAN - **O**il **N**atural **A**ir **N**atural

A transformer with a liquid coolant with a flashpoint less than or equal to 300°C (oil) that is naturally flowing through the radiator fins from thermal convection, with outside ambient air naturally flowing through the radiator fins without assistance.

The answer is: **See completed drag and drop diagram above.**

(Drag and drop AIT question) - Ch. 7.1 Transformers

Problem #42 Solution

Reactors are large inductors sized to handle higher voltage and current in power systems. Let's evaluate each of the possible answer choices to see which ones are appropriate applications for reactors used in power systems:

(A) Transient current prevention.

When current flows through an inductor, a strong magnetic field is created that limits the rate of change of current flowing through it:

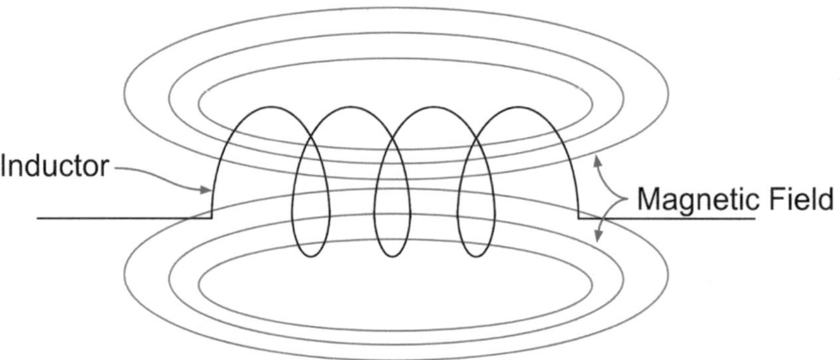

Because of this, reactors are often used in power systems to prevent equipment damage by reducing current peaks during transient overcurrent events such as those created by lightning strikes.

True! **Reactors can be used to prevent transient current.**

(B) Series transmission line compensation.

Since current carrying conductors are inductive, using a reactor for series line compensation would only increase the overall inductance of the circuit which would make the overall power factor worse instead of better.

The goal of series line compensation is to improve the circuit power factor by reducing the inductance of the circuit and is mostly achieved by using capacitors, not reactors.

False! **Reactors are not suitable for series transmission line compensation.**

(C) Voltage support to increase load voltage.

False! **Shunt capacitor banks, not reactors, are used to increase voltage, mostly on the secondary side of power transformers.**

(D) Harmonic reduction.

Similar to how reactors can be used to prevent transient overcurrent, they can also be used to filter harmonic currents by reducing the additive and subtractive effects that they have on the fundamental 60 Hz power supply by limiting the rate of change of all non-fundamental 60 Hz current peaks.

True! **Reactors can be used for harmonic reduction.**

(E) Prevent damage to motors fed from variable-frequency drives.

The output waveform of a variable-frequency drive (VFD) is a pulse width modulated (PWM) signal which is not a true sine wave. This results in current spikes drawn by motor loads, especially when the motor is located far from the VFD compared to what would normally be a sinusoidal current.

Reactors placed between a variable-frequency drive and a motor help to smooth out the current drawn by the motor by limiting the rate of change flowing through the reactor. Since current spikes damage insulation, reactors help reduce the damage of motors fed from variable-frequency drives.

True! **Reactors help smooth out the current drawn by motors from VFDs that would otherwise lead to damage.**

The answer is: **A, D, and E**

(Multiple correct AIT question) - Ch. 7.2 Reactors

Problem #43 Solution

The transformer short circuit test[59] is used to determine the winding resistance (R_{eq}) and leakage (X_{eq}) reactance of the transformer. It can also be used to calculate the I^2R copper winding losses.

The test is typically carried out by the manufacturer so that this data can be included on the transformer nameplate and given to the customer:

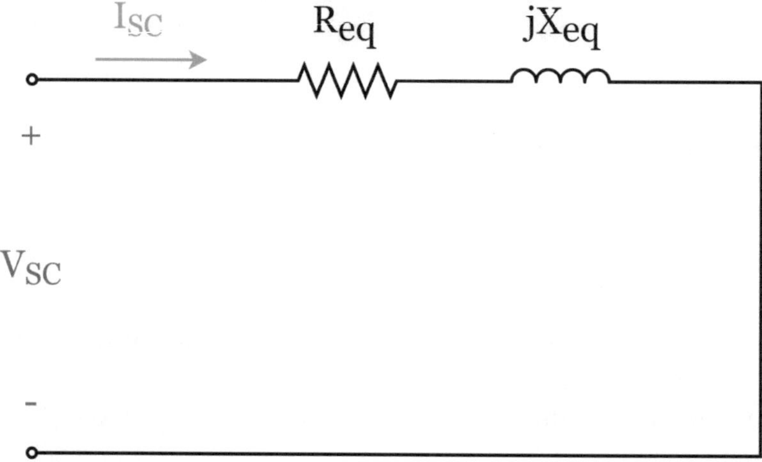

This information is often represented as the **winding impedance** in ohms ($Z_{eq} = R_{eq} + X_{eq}$) or as the **percent impedance** (%Z).

The answer is: **B and C.**

(Multiple correct AIT question) - Ch. 7.3 Testing

[59] *NCEES® Reference Handbook (Version 1.1.2) - 4.3.3.2 Short-Circuit Test p. 64*

Problem #44 Solution

Let's fill in the transformer short circuit test[60] equivalent circuit using the information given in the problem:

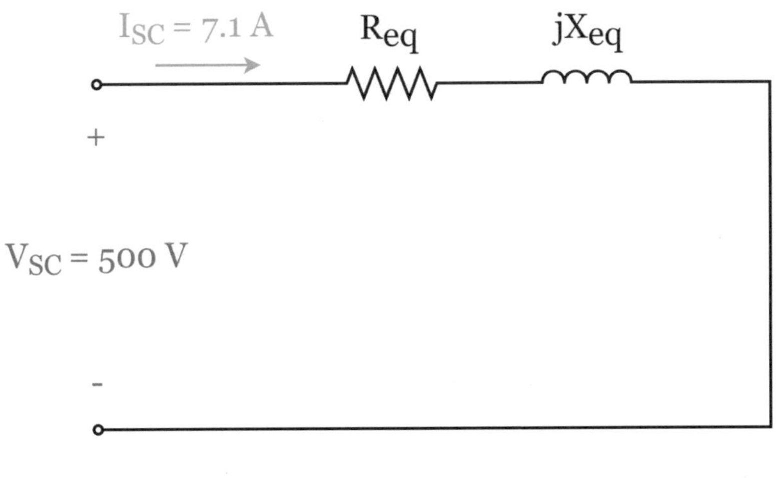

PSC = 1,230 W

First, let's calculate the transformer impedance magnitude $|Z_{eq}|$ using Ohm's law:

$$|Z_{eq}| = \frac{V_{SC}}{I_{SC}}$$
$$|Z_{eq}| = \frac{500\ V}{7.1\ A}$$

```
DEG
500/7.1
          70.42253521
```

The transformer impedance magnitude is approximately 70.4 Ω.

Next, let's calculate the transformer base impedance[61] using the transformer ratings given in the problem as the base values in the per unit system.

Since the short circuit measurements were taken on the high voltage side of the single-phase transformer, we'll need to use the transformer high voltage rating for the base voltage:

[60] *NCEES® Reference Handbook (Version 1.1.2) - 4.3.3.2 Short-Circuit Test p. 64*
[61] *NCEES® Reference Handbook (Version 1.1.2) - 3.1.3 Per Unit System p. 34*

$$Z_B = \frac{V_B^2}{S_B}$$

$$Z_B = \frac{15\,kV^2}{112\,kVA}$$

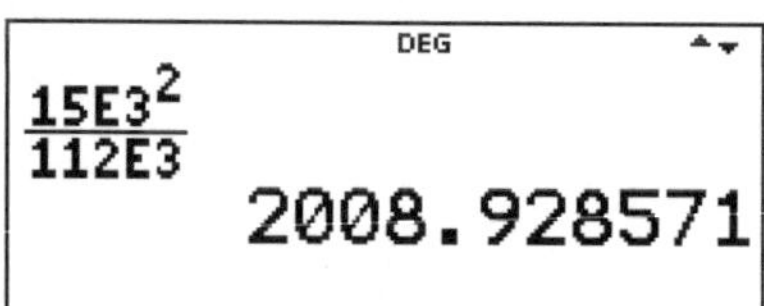

The transformer base impedance is approximately 2,008.9 Ω.

The last step is to calculate the percent impedance by calculating the per unit impedance of the transformer and converting to a percentage:

$$Z_{PU} = \frac{Z}{Z_B}$$

$$Z_{PU} = \frac{70.4\,\Omega}{2,008.9\,\Omega}$$

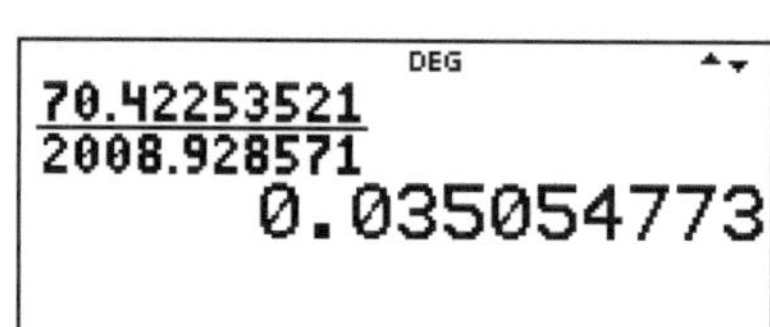

The transformer per unit impedance is approximately 0.035 pu.

Note: Since this is a fill in the blank AIT problem, we'll want to use the exact values in the calculator by scrolling up with the calculator arrows and pressing enter instead of the approximate values since that can affect the rounding of the final answer.

Converted to a percentage and rounded to one decimal place, the percent impedance of the transformer is 3.5%.

The answer is: **3.5%**

(Fill in the blank AIT question) - Ch. 7.3 Testing

Problem #45 Solution

The transformer open circuit test[62] is used to determine the **core loss resistance (R_C)** and the **magnetizing reactance (X_M)** of the transformer. It can also be used to calculate the no load core losses of the transformer.

Just like the short circuit test, the open circuit test is typically carried out by the manufacturer so that this data can be included on the transformer nameplate and given to the customer:

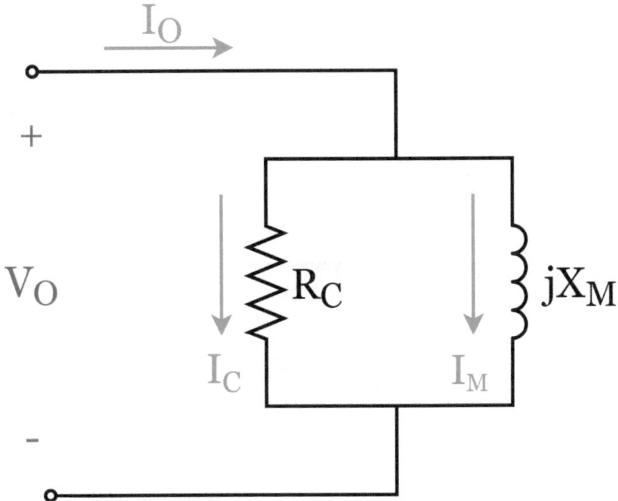

The answer is: **D and E.**

(Multiple correct AIT question) - Ch. 7.3 Testing

[62] *NCEES® Reference Handbook (Version 1.1.2) - 4.3.3.1 Open-Circuit Testing p. 64*

Problem #46 Solution

Let's fill in the transformer no load circuit test[63] (also known as the open circuit test) equivalent circuit using the information given in the problem:

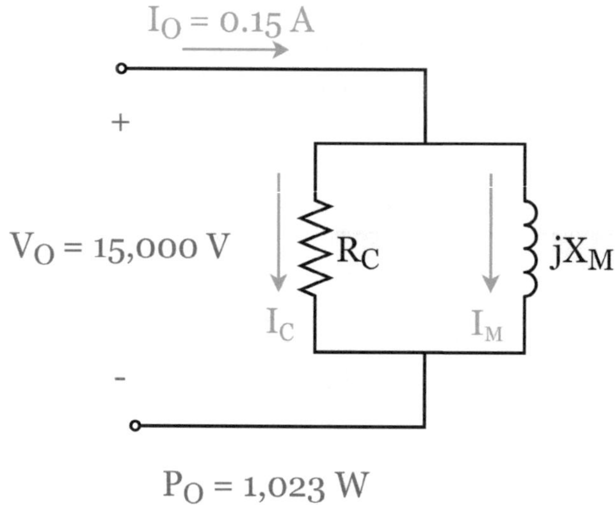

First, let's calculate the angle of the no load open circuit current (I_o):

$$\cos(\theta_o) = \frac{P_o}{V_o I_o}$$

$$\theta_o = \cos^{-1}\left(\frac{P_o}{V_o I_o}\right)$$

$$\theta_o = \cos^{-1}\left(\frac{1,023\,W}{15,000\,V \cdot 0.15\,A}\right)$$

```
DEG
cos-1( 1023 / 15000×0.15
                62.95651097
```

The angle of the no load open circuit current (I_o) is approximately 63 degrees.

[63] *NCEES® Reference Handbook (Version 1.1.2) - 4.3.3.1 Open-Circuit Testing p. 63*

Next, let's use the angle of the no load open circuit current (θ_o) along with the magnitude of the open circuit current (I_O) from the test data given in the problem to calculate the magnetizing current (I_M):

$$I_m = I_o \sin(\theta_o)$$
$$I_m = 0.15\,A \cdot \sin(63)$$

DEG
0.15sin(62.9565▸
0.133599251

The magnetizing current (I_M) is approximately 0.134 amps.

Note: Since this is a fill in the blank AIT problem, we'll want to use the exact values in the calculator by scrolling up with the calculator arrows and pressing enter instead of the approximate values since that can affect the rounding of the final answer.

The last step is to use Ohm's law to calculate the magnetizing reactance (X_M) from the magnetizing current (I_M) and the open circuit voltage (V_O) given in the problem's test data:

$$X_m = \frac{V_o}{I_m}$$
$$X_m = \frac{15,000\,V}{0.134\,A}$$

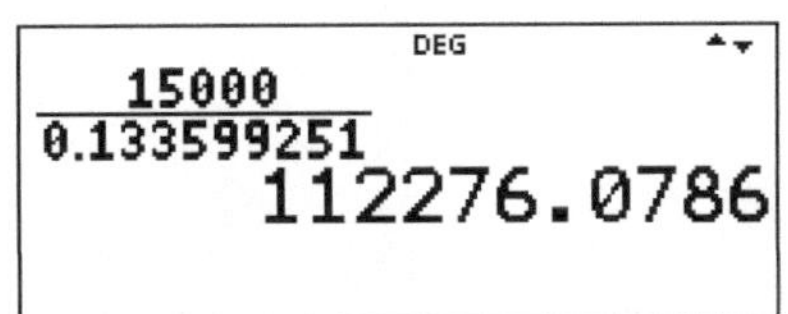

The magnetizing reactance (X_M) is approximately 112,267 ohms.

Rounded to the nearest kiloohm, the magnetizing reactance is approximately 112 kΩ.

The answer is: **112**

(Fill in the blank AIT question) - Ch. 7.3 Testing

Problem #47 Solution

Let's evaluate each of the possible answer choices to see which ones are appropriate applications for capacitors[64] in power systems:

(A) Power factor correction

One of the most common applications for capacitors in power systems is power factor correction. Capacitors supply reactive power and can be used to improve the overall power factor of the system they are connected to.

The primary use of capacitors for power factor correction is to decrease or prevent fees the electric utility charges to customers with poor power factors.

True! **Capacitors are used for power factor correction.**

(B) Voltage support

By reducing the reactive power flow in power systems, shunted capacitors are able to decrease voltage drop and help maintain nominal voltage levels.

This application is typically utilized on the secondary side of service transformers or at distribution voltage buses where the voltage level is too low to meet the needs of the connected load.

True! **Capacitors are used for voltage support.**

(C) Transmission line series compensation

Transmission line conductors with a large inductive reactance will consume large amounts of reactive power and result in a large voltage drop. Series compensators made of capacitors can be used to reduce the overall inductive reactance of the transmission line.

The end result is a decrease in the amount of reactive power consumed in the transmission line conductor and a decrease in voltage drop.

True! **Capacitors are used for series compensation.**

64 *NCEES® Reference Handbook (Version 1.1.2) - 4.3.4 Capacitors p. 64*

(D) Harmonic filter

Harmonic filters block or shunt to ground harmonic current in power systems. Most harmonic current comes from nonlinear loads such as power electronics and switching circuits.

Harmonics can cause undesirable effects on power systems such as an increase in heating, an increase in current (especially on the neutral), and weakened magnetic fields in electric power devices such as transformers and rotating machines. Harmonic filters are mostly composed of capacitors and inductors.

True! **Capacitors are used for harmonic filters.**

(E) Ground fault protection

Ground fault protection is a protection scheme that monitors the ground fault return path for unintended faults to ground that can result in dangerous levels of ground fault current. Ground fault current can pose a safety threat to personnel and can damage equipment.

Ground fault protection is typically accomplished using current transformers (CTs), protective relays, and overcurrent protection devices (OCPDs). They do not use capacitors.

False! **Capacitors are not used for ground fault protection.**

The answer is: **A, B, C, and D.**

(Multiple correct AIT question) - Ch. 7.4 Capacitors

Problem #48 Solution

Let's evaluate each statement to see which ones are true if n equal conductors are connected in parallel for each phase in order to supply power to a load:

(A) The voltage drop across each phase will increase by a factor of n.

According to Ohm's law, voltage drop (V) is directly proportional to impedance (Z) if the current (I) is constant:

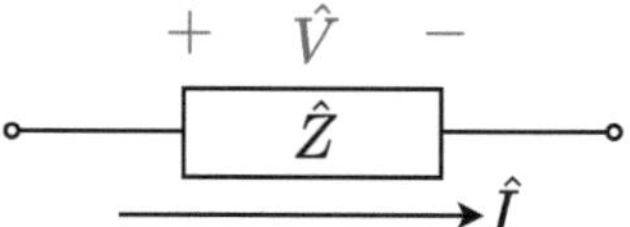

$$\hat{V} = \hat{I}\,\hat{Z}$$

Before we can use Ohm's law to determine how voltage drop is affected by parallel conductors, we have to first determine how connecting equal conductors in parallel changes the overall impedance of each phase compared to a single individual conductor.

We can do this by calculating the equivalent impedance (Z_{eq}) of n conductors in parallel by adding the impedance of each conductor using the reciprocal of reciprocal sums method:

$$\hat{Z}_{eq} = \hat{Z}_1 // \hat{Z}_2 // \hat{Z}_3 // \ldots // \hat{Z}_n$$

$$\hat{Z}_{eq} = \frac{1}{\frac{1}{\hat{Z}_1} + \frac{1}{\hat{Z}_2} + \frac{1}{\hat{Z}_3} + \ldots + \frac{1}{\hat{Z}_n}}$$

If each conductor that is connected in parallel is the same, then the impedance of each conductor is equal ($\hat{Z} = \hat{Z}_1 = \hat{Z}_2 = \hat{Z}_3 = \ldots = \hat{Z}_n$) and the above formula simplifies to:

$$\hat{Z}_{eq} = \frac{\hat{Z}}{n}$$

where:

Z_{eq} = Equivalent parallel impedance.
Z = The individual impedance of each equal conductor connected in parallel.
n = The number of equal conductors connected in parallel.

It's easier to understand this relationship by using a definite value for n.

Let's try it by determining the equivalent impedance (Z_{eq}) of two (n = 2) equal conductors connected in parallel:

$$\hat{Z}_{eq} = \hat{Z}//\hat{Z}$$

$$\hat{Z}_{eq} = \frac{1}{\frac{1}{\hat{Z}} + \frac{1}{\hat{Z}}}$$

$$\hat{Z}_{eq} = \frac{1}{\frac{2}{\hat{Z}}}$$

$$\hat{Z}_{eq} = \frac{\hat{Z}}{2}$$

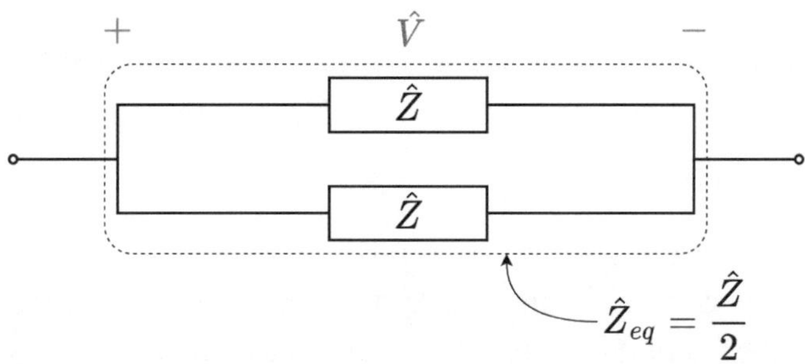

The equivalent impedance (Z_{eq}) of two (n = 2) equal conductors connected in parallel is equal to the individual impedance of each equal conductor connected in parallel (Z) divided by the number of conductors connected in parallel.

Let's try it one more time by determining the equivalent impedance (Z_{eq}) of three (n = 3) equal conductors connected in parallel:

$$\hat{Z}_{eq} = \hat{Z}//\hat{Z}//\hat{Z}$$

$$\hat{Z}_{eq} = \frac{1}{\frac{1}{\hat{Z}} + \frac{1}{\hat{Z}} + \frac{1}{\hat{Z}}}$$

$$\hat{Z}_{eq} = \frac{1}{\frac{3}{\hat{Z}}}$$

$$\hat{Z}_{eq} = \frac{\hat{Z}}{3}$$

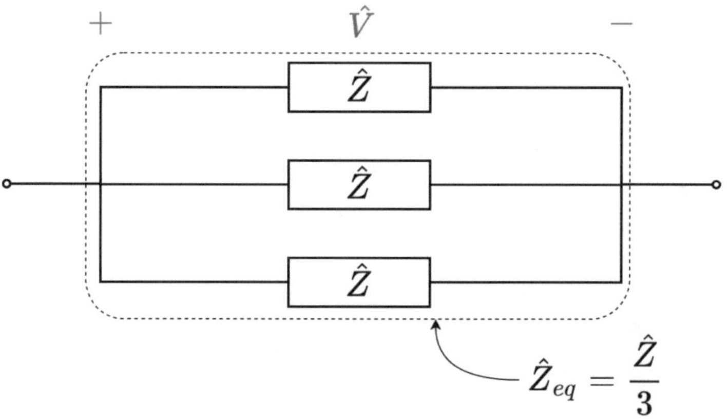

Again we find that the equivalent impedance (Z_{eq}) of three (n = 3) equal conductors connected in parallel is equal to the individual impedance of each conductor (Z) divided by the number of conductors.

Now that we have a clear understanding of how equal conductors connected in parallel affect the overall phase impedance, let's plug this relationship back into Ohm's law to determine the effect that n equal parallel conductors have on the voltage drop for each phase.

The voltage drop (V) for a single conductor that has an impedance of Z:

$$\hat{V} = \hat{I}\,\hat{Z}$$

The voltage drop (V) for n equal conductors connected in parallel, where each individual conductor has an impedance of Z:

$$\hat{V} = \hat{I}\,\frac{\hat{Z}}{n}$$

Looking at the relationship above, n equal conductors connected in parallel will **decrease** the voltage drop (V) across each phase by the same factor of n, **not** increase it.

False! **The voltage drop across each phase will NOT increase by a factor of n.**

(B) The voltage drop across each phase will decrease by a factor of n.

We discovered that this is true according to Ohm's law while evaluating statement (A):

$$\hat{V} = \hat{I}\,\frac{\hat{Z}}{n}$$

where:

V = Voltage drop across each phase.
Z = The individual impedance of each equal conductor connected in parallel.
n = The number of equal conductors connected in parallel.

True! **The voltage drop across each phase will decrease by a factor of n.**

(C) The current flowing through each conductor will increase by a factor of n.

Parallel conductors act as a current divider by providing multiple paths for the current to flow through on its way from the source to the load.

First, let's start with the relationship for the voltage drop across each phase with n parallel conductors that we arrived at while evaluating statement (A):

$$\hat{V} = \hat{I}\,\frac{\hat{Z}}{n}$$

We can calculate the current flowing through each individual conductor in the bundle (I_Z) using Ohm's law:

$$\hat{I}_Z = \frac{\hat{V}}{\hat{Z}}$$

Next, substitute in the expression for the voltage drop (V) across each phase with n equal parallel conductors and simplify:

$$\hat{I}_Z = \frac{\hat{I}\frac{\hat{Z}}{n}}{\hat{Z}}$$

$$\hat{I}_Z = \frac{\hat{I}}{n}$$

The current that would normally flow through just one phase conductor (I) has now been decreased by a factor of n:

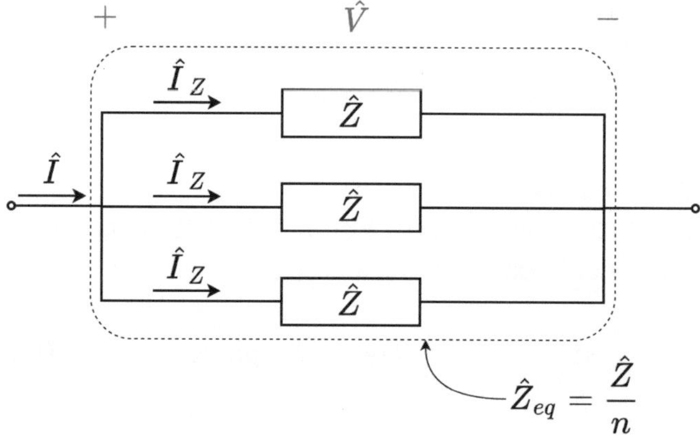

False! **The current flowing through each conductor will NOT increase by a factor of n.**

(D) The current flowing through each conductor will decrease by a factor of n.

We found this to be true while evaluating the previous statement, (C):

$$\hat{I}_Z = \frac{\hat{I}}{n}$$

True! **The current flowing through each conductor will decrease by a factor of n.**

(E) The equivalent line impedance for each phase will decrease by a factor of n.

We found this to be true while evaluating statement (A):

$$\hat{Z}_{eq} = \frac{\hat{Z}}{n}$$

Where:

Z_{eq} = Equivalent parallel impedance.
Z = The individual impedance of each equal conductor connected in parallel.
n = The number of equal conductors connected in parallel.

True! **The equivalent line impedance for each phase will decrease by a factor of n.**

The answer is: **B, D, and E.**

(Multiple correct AIT question) - Ch. 8.1 Voltage Drop

Problem #49 Solution

Let's evaluate each possible statement to identify which ones would improve the voltage regulation of a short transmission line with rated voltage at the sending side and a constant load current.

(A) Decreasing the series line impedance.

Let's look at the short transmission line single-phase equivalent circuit[65]:

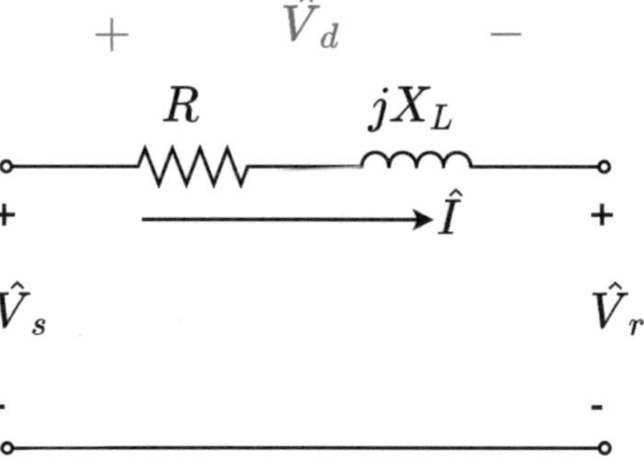

As the load current (I) flows from the sending side to the receiving side, a voltage drop (V_d) is created across the series line impedance ($Z = R+jX_L$) according to Ohm's law:

$$\hat{V}_d = \hat{I}\,\hat{Z}$$

Notice that the voltage drop (V_d) will **decrease** linearly along with any decrease in the series line impedance (Z):

$$\downarrow \hat{V}_d = \hat{I}\left(\downarrow \hat{Z}\right)$$

A decreased voltage drop (V_d) will result in an **increase** in the available voltage at the receiving end (V_r) according to Kirchhoff's voltage law (KVL):

$$\hat{V}_r = \hat{V}_s - \hat{V}_d$$

$$\uparrow \hat{V}_r = \hat{V}_s - \downarrow \hat{V}_d$$

Now, let's evaluate how this affects the voltage regulation[66] of the short transmission line:

---

[65] NCEES® Reference Handbook (Version 1.1.2) - 5.1.6.1 Short Transmission Line Model p.72
[66] NCEES® Reference Handbook (Version 1.1.2) - 5.1.6 Transmission Line Models p. 71

$$\%VR = \frac{|V_{r_{nl}}| - |V_{r_{fl}}|}{|V_{r_{fl}}|} \times 100$$

where:

$|V_{r\,nl}|$ = Magnitude of the receiving voltage under **no load** conditions.
$|V_{r\,fl}|$ = Magnitude of the receiving voltage under **full load** conditions.

For a short transmission line, the no load ($I = o$) receiving voltage ($V_{r\,nl}$) is equal to the sending voltage (V_s) since there is no voltage drop across the line impedance (Z):

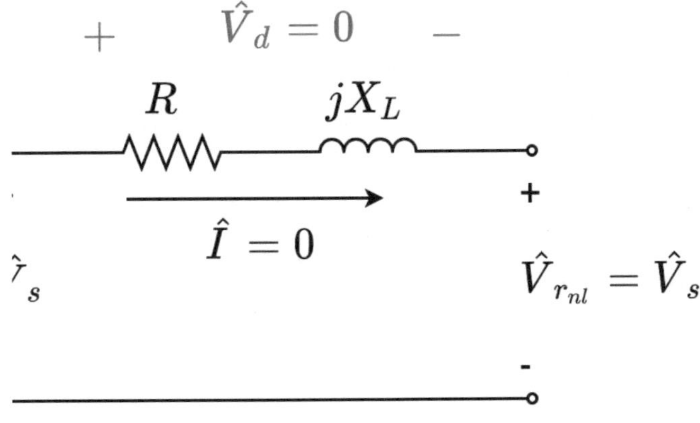

No load conditions

At no load conditions, the load current is equal to zero:

$$\hat{I} = 0$$

This results in no voltage drop (V_d) across the series line impedance (Z):

$$\hat{V}_d = \hat{I}\,\hat{Z}$$
$$\hat{V}_d = (0)\hat{Z}$$
$$\hat{V}_d = 0$$

which results in the no load receiving voltage ($V_{r\,nl}$) being equal to the sending voltage (V_s):

$$\hat{V}_r = \hat{V}_s - \hat{V}_d$$
$$\hat{V}_{r\,nl} = \hat{V}_s - 0$$

$$\hat{V}_{r\,nl} = \hat{V}_s$$

For a short transmission line, the voltage regulation[67] formula simplifies to:

$$\%VR = \frac{|V_{r_{nl}}| - |V_{r_{fl}}|}{|V_{r_{fl}}|} \times 100$$

$$\%VR = \frac{|V_s| - |V_r|}{|V_r|} \times 100$$

where:

$|V_s|$ = Magnitude of the sending voltage.
$|V_r|$ = Magnitude of the receiving voltage solved for when $I \neq 0$.

Notice that as the available voltage at the receiving end (V_r) **increases**, the voltage regulation **decreases**:

$$\downarrow \%VR = \frac{|V_s| - \uparrow |V_r|}{\uparrow |V_r|} \times 100$$

A **decrease** in percent voltage regulation results in **improved** voltage regulation since lower voltage regulation is more desirable and reflects the transmission line's ability to maintain the available voltage at the receiving end (V_r) to be closer in value to the voltage available at the sending end (V_s).

True! **Decreasing the series line impedance will improve the voltage regulation for a short transmission line.**

(B) Decreasing the shunt line capacitance.

While the medium[68] and long[69] transmission line circuit model includes the shunt line capacitance using the "pi model," the short transmission line module does not.

The effects of shunt line capacitance on a transmission line 50 miles or less in length is negligible and therefore is not included in the short transmission line module:

[67] NCEES® Reference Handbook (Version 1.1.2) - 5.1.6 Transmission Line Models p. 71
[68] NCEES® Reference Handbook (Version 1.1.2) - 5.1.6.2 Medium Transmission Line Model p. 73
[69] NCEES® Reference Handbook (Version 1.1.2) - 5.1.6.3 Long Transmission Line Model p. 73

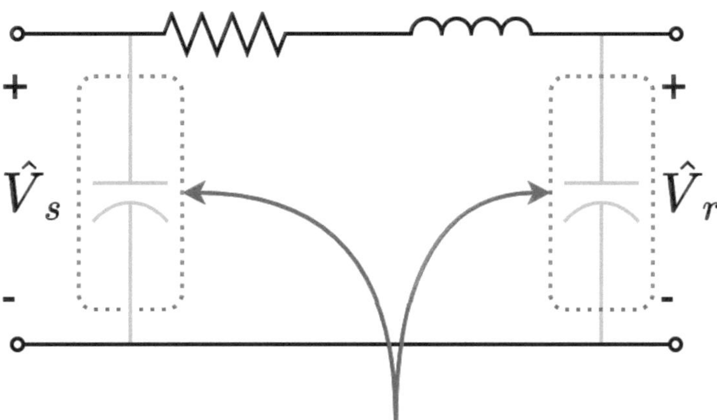

Since shunt line capacitance is neglected in the short transmission line module, it does not affect the available voltage at the receiving end (V_r), which means it does not have an effect on voltage regulation.

> *Note: this is **only** true for the short transmission line model and **not** the medium or long transmission line models.*

False! **Decreasing the shunt line capacitance does not improve voltage regulation for a short transmission line.**

(C) Improve the load power factor.

As the load power factor **improves** (increases), the load current phase angle (θ_I) will **increase** (rotate counterclockwise in the positive angle direction) and approach the value of the receiving voltage phase angle (θ_{Vr}):

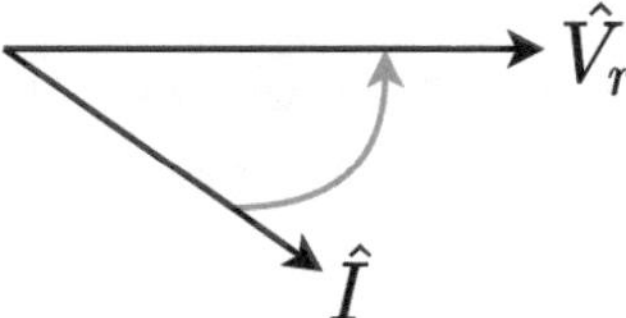

This will result in a **decreased** phase angle for the voltage drop (V_d) according to Ohm's law:

$$\hat{V}_d = \hat{I}\,\hat{Z}$$

As the voltage drop (V_d) phase angle **decreases**, the magnitude of the voltage at the receiving end (V_r) will **increase** according to Kirchhoff's voltage law (KVL):

$$\hat{V}_r = \hat{V}_s - \hat{V}_d$$

As the magnitude of voltage at the receiving end (V_r) **increases**, the voltage regulation will **improve** (decrease), just as we learned earlier while evaluating statement (A) of this problem:

$$\downarrow \%VR = \frac{|V_s| - \uparrow |V_r|}{\uparrow |V_r|} \times 100$$

True! **Improving the load power factor will improve the voltage regulation for a short transmission line.**

(D) Install a series reactive line compensator.

A series reactive line compensator is a capacitive reactance (X_c) connected in series with a transmission line conductor:

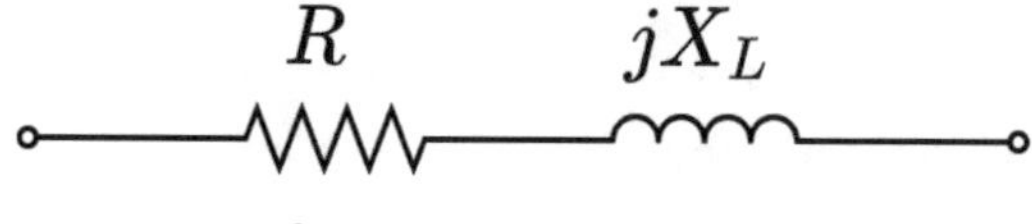

Before series compensation

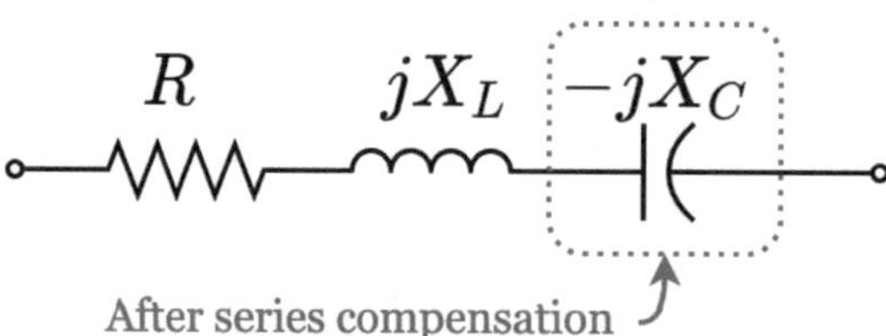

After series compensation

Because capacitive reactance (X_c) is **negative**, it **decreases** the overall reactance (X) of the transmission line which, as a result, **decreases** the series transmission line impedance (Z):

$$\hat{Z} = R + jX_L - jX_C$$
$$\hat{Z} = R + j(X_L - X_C)$$

Just as we learned while evaluating statement (A) of this problem earlier, as the series transmission line impedance (Z) **decreases**, the following occurs:

- The voltage drop (V_d) will **decrease.**
- The available voltage at the receiving end (V_r) will **increase.**
- The voltage regulation will **improve (decrease).**

True! **Installing a series reactive line compensator will improve the voltage regulation for a short transmission line.**

(E) Install a recloser.

A recloser is a type of overcurrent protection device (OCPD) designed to open a circuit during fault conditions and then wait a predetermined amount of time in the range of 0 to 10 seconds before attempting to reclose the circuit.

The goal of a recloser is to restore power to a circuit during transient fault conditions, like lightning strikes, that clear up on their own.

Reclosers have no impact on the voltage regulation of a transmission line.

False! **Installing a recloser does not improve voltage regulation for a short transmission line.**

The answer is: **A, C, and D.**

(Multiple correct AIT question) - Ch. 8.2 Voltage Regulation

Problem #50 Solution

Let's evaluate each possible statement to identify which ones are true for a three-phase wye connected capacitor bank that is reassembled into a delta connected capacitor bank.

(A) The voltage applied across each phase of the capacitor bank will increase by a factor of the square root of three.

The voltage across each phase of a **wye** connection is the phase voltage (V_P) of the system[70], also known as the system line-to-neutral voltage:

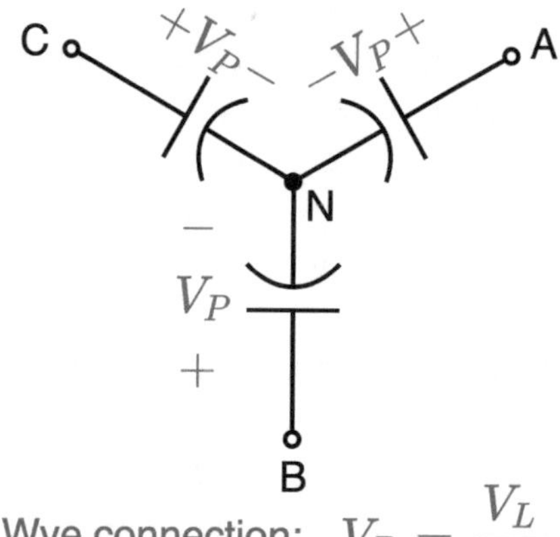

Wye connection: $V_P = \dfrac{V_L}{\sqrt{3}}$

If we take the same wye connected capacitor bank and reassemble it into a **delta** connected capacitor bank, then the voltage across each phase will now be equal to the line voltage (V_L) of the system[71] since for a delta connection, phase voltage is equal to line voltage ($V_P = V_L$):

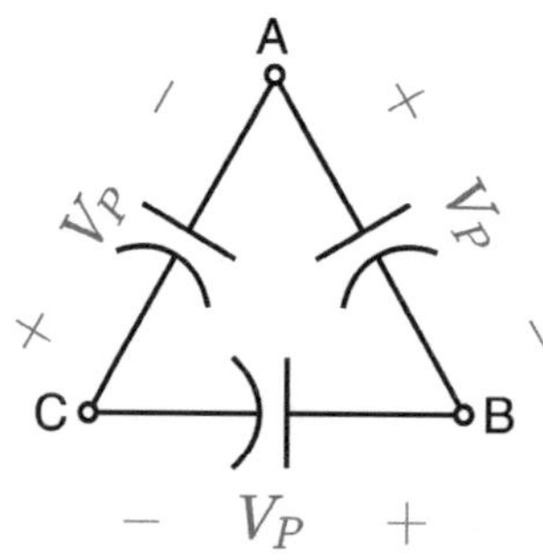

Delta connection: $V_P = V_L$

Notice that the voltage applied across each phase of the **wye** connected capacitor bank is **smaller** by a factor of the square root of three compared to the **delta** connected capacitor bank.

---

[70] NCEES® Reference Handbook (Version 1.1.2) - 3.1.1 3 Phase Circuits p. 33
[71] NCEES® Reference Handbook (Version 1.1.2) - 3.1.1 3 Phase Circuits p. 33

Looking at this relationship from the other perspective, we can say that the voltage applied across each phase of the **delta** connected capacitor bank is **larger** by a factor of the square root of three compared to the **wye** connected capacitor bank:

$$V_P = \frac{V_L}{\sqrt{3}}$$

Wye connected

$$V_P = V_L$$

Delta connected

True! **The voltage applied across each phase of the capacitor bank will increase by a factor of the square root of three.**

(B) The reactive power supplied by the capacitor bank will increase by a factor of three.

The total reactive power ($Q_{3\varnothing}$) supplied by a three-phase capacitor bank is:

$$Q_{3\varnothing} = 3\frac{|V_P|^2}{X_C}$$

where:

$|V_p|$ = The magnitude of the voltage across each phase of the capacitor bank.
X_C = The per-phase capacitive reactance of the capacitor bank.

Note: this formula is not currently in the Reference Handbook, but if you've attended our Live Class, it should be very familiar to you.

When the capacitor bank is **wye** connected, the phase voltage (V_P) across each phase is the line-to-neutral voltage of the three-phase system:

$$V_P = \frac{V_L}{\sqrt{3}}$$

When we substitute this relationship into the reactive power ($Q_{3\varnothing}$) formula above, the coefficient of three (3) cancels out:

$$Q_{3\varnothing} = 3\frac{|V_P|^2}{X_C}$$

$$Q_{3ø} = 3\frac{\left|\frac{V_L}{\sqrt{3}}\right|^2}{X_C}$$

$$Q_{3ø} = \frac{3}{3}\frac{|V_L|^2}{X_C}$$

$$Q_{3ø} = \frac{|V_L|^2}{X_C}$$

When the capacitor bank is **delta** connected, the phase voltage (V_P) across each phase is the line voltage of the three-phase system:

$$V_P = V_L$$

When we substitute this relationship into the reactive power ($Q_{3ø}$) formula above, the line voltage magnitude replaces the phase voltage magnitude:

$$Q_{3ø} = 3\frac{|V_P|^2}{X_C}$$

$$Q_{3ø} = 3\frac{|V_L|^2}{X_C}$$

Notice that when we compare the final reactive power ($Q_{3ø}$) formula for the **delta** connected capacitor bank to the **wye** connected capacitor bank, the **delta** connected capacitor bank will supply three times as much reactive power:

$$Q_{3ø} = \frac{|V_L|^2}{X_C}$$

Wye connected

$$Q_{3ø} = 3\frac{|V_L|^2}{X_C}$$

Delta connected

True! **The reactive power supplied by the capacitor bank will increase by a factor of three.**

(C) The line charging current drawn by the capacitor bank will increase by a factor of three.

We can calculate line current (I_L) drawn by the capacitor bank using three-phase reactive power ($Q_{3ø}$) and line voltage (V_L)[72]:

$$|I_L| = \frac{Q_{3ø}}{\sqrt{3}|V_L|}$$

Note: this is the standard three-phase apparent power formula ($S_{3ø}$) using reactive power instead since a capacitor bank will only supply reactive power (S = Q < 90ºFor a capacitor bank).

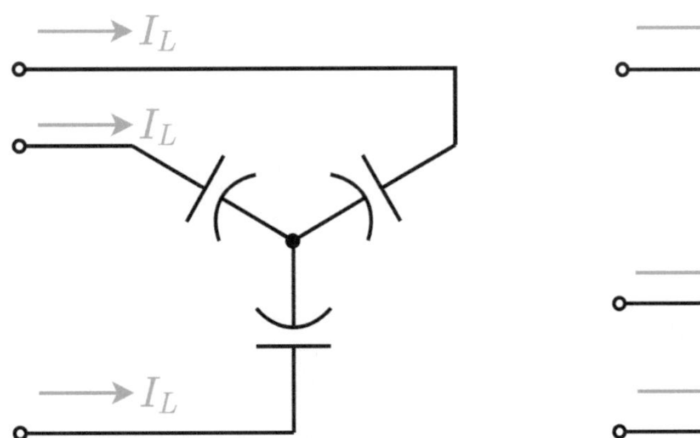

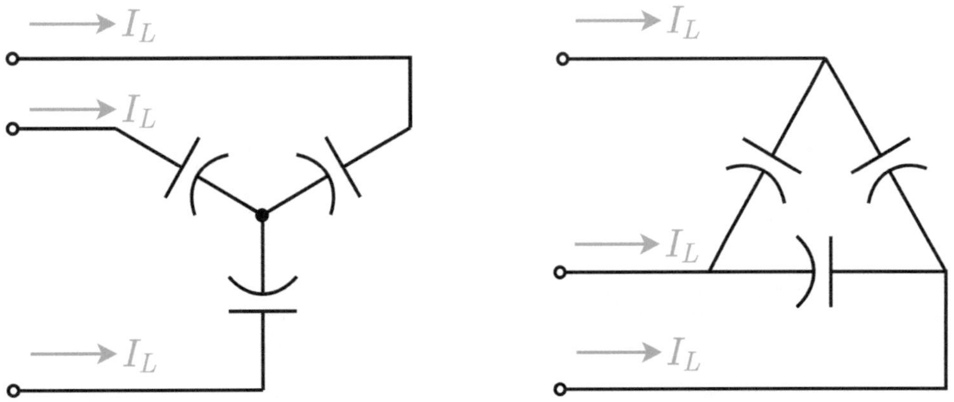

Since the capacitor bank will supply three times as much reactive power when it is **delta** connected, as we learned while evaluating the previous statement (B), then the line charging current drawn by the **delta** capacitor bank will also be three times as large compared to when it is wye connected.

Wye connected:

$$|I_L| = \frac{Q_{3ø}}{\sqrt{3}|V_L|}$$

$$Q_{3ø} = \frac{|V_L|^2}{X_C}$$

$$|I_L| = \frac{|V_L|}{\sqrt{3}X_C}$$

Delta connected:

$$|I_L| = \frac{Q_{3ø}}{\sqrt{3}|V_L|}$$

$$Q_{3ø} = 3\frac{|V_L|^2}{X_C}$$

$$|I_L| = 3\frac{|V_L|}{\sqrt{3}X_C}$$

72 NCEES® Reference Handbook (Version 1.1.2) - 3.1.1 3 Phase Circuits p. 33

True! **The line charging current drawn by the capacitor bank will increase by a factor of three.**

(D) The reactance of each phase of the capacitor bank will increase by a factor of three.

According to the problem, a three-phase power factor correction wye connected capacitor bank is reassembled into a delta connected capacitor bank.

No change is being made to the amount of reactance (X_C) in each phase; the connections between each phase are being reassembled to change the connection type from wye to delta:

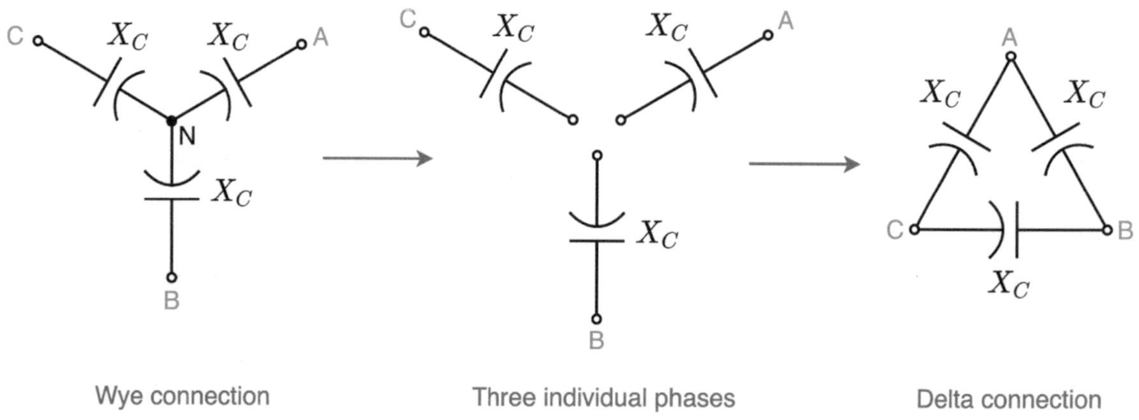

Wye connection　　Three individual phases　　Delta connection

False! **The reactance of each phase of the capacitor bank will NOT increase by a factor of three.**

(E) The capacitance of each phase of the capacitor bank will increase by a factor of three.

Capacitance (C) and capacitive reactance (X_C) are two different ways of expressing the value of a capacitor. Capacitance (C) is in the unit of farad [F], which is frequency **independent**, and capacitive reactance (X_C) is in the unit of ohm [Ω], which is frequency **dependent**.

We can convert from capacitive reactance (X_C) to capacitance (C) using the following formula[73]:

[73] NCEES® Reference Handbook (Version 1.1.2) - 3.1.4.1 AC Circuits p. 36

$$C = -\frac{1}{wX_C}$$

where:

w = Angular velocity ($w = 2\pi f$).

Notice in the formula above that capacitance (C) will remain unchanged if capacitive reactance (X_C) does not change when the three-phase capacitor bank is reassembled from a wye connection to a delta connection.

False! **The capacitance of each phase of the capacitor bank will NOT increase by a factor of three.**

The answer is: **A, B, and C.**

(Multiple correct AIT question) - Ch. 8.3 Power Factor Correction and Voltage Support

Problem #51 Solution

During a fault, the voltage measured on each faulted phase will **decrease** to as low as zero volts depending on the value of the fault impedance.

The current measured on each faulted phase will dramatically **increase,** compared to the pre-faulted load current, due to the presence of short circuit current.

Single-phase to ground fault:

The first three-phase voltage waveform shows a **single-phase to ground fault**, also known as a single-line to ground fault, or a single-phase fault.

Notice only the voltage on one phase has decreased at the time of the fault. In this particular example, this is a C-phase to ground fault:

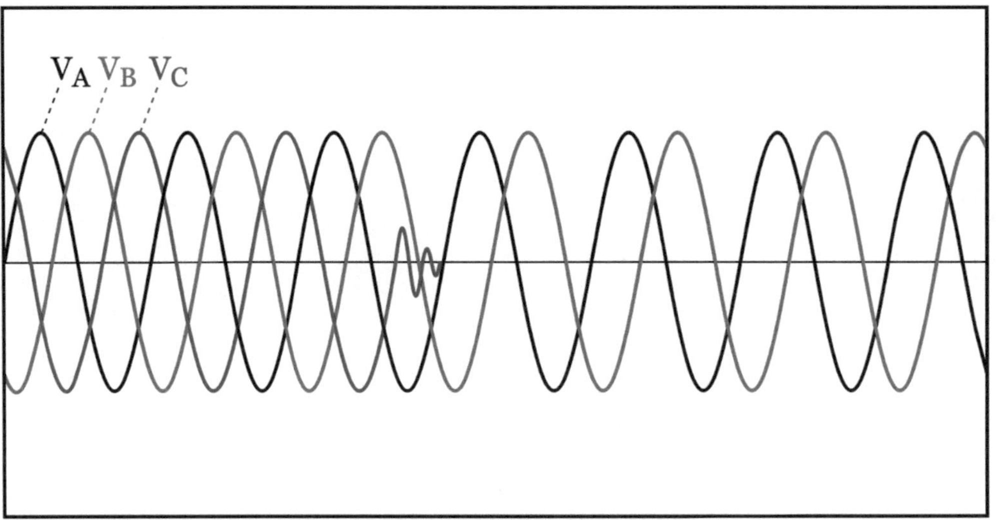

Single-phase to Ground Fault

If we were looking at the current waveform for this type of fault, the current on C-phase would have dramatically increased at the same time when the C-phase voltage decreased.

Three-Phase Fault:

The second three-phase voltage waveform shows a **three-phase fault**, also known as a line to line to line fault.

Notice the voltage on all three phases has decreased at the time of the fault:

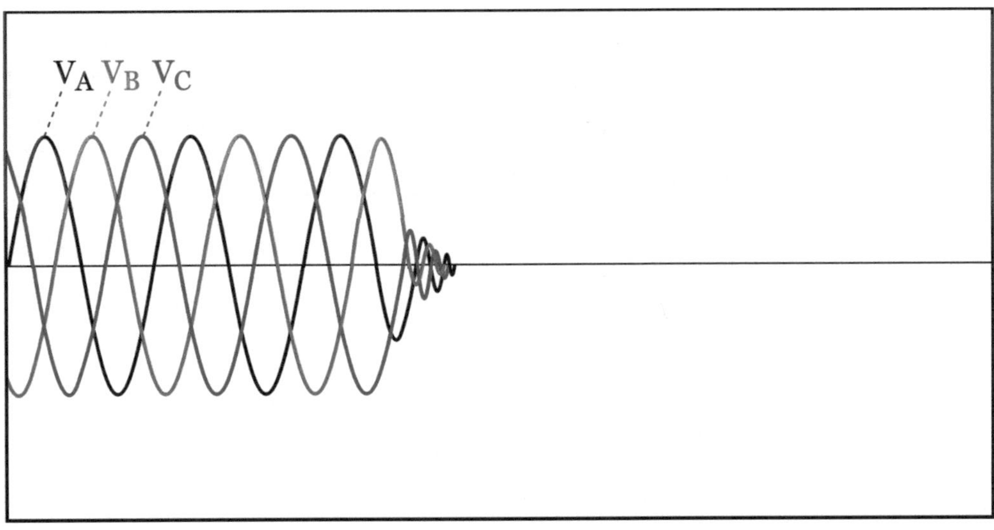

Three Phase Fault

If we were looking at the current waveform for this type of fault, the current on all three phases would have dramatically increased at the same time when voltages decreased.

Normal Operating Conditions:

The third three-phase voltage waveform shows normal operating conditions. There is no current fault present.

Notice the voltage on all three phases remains at nominal magnitudes displaced evenly 120 degrees apart:

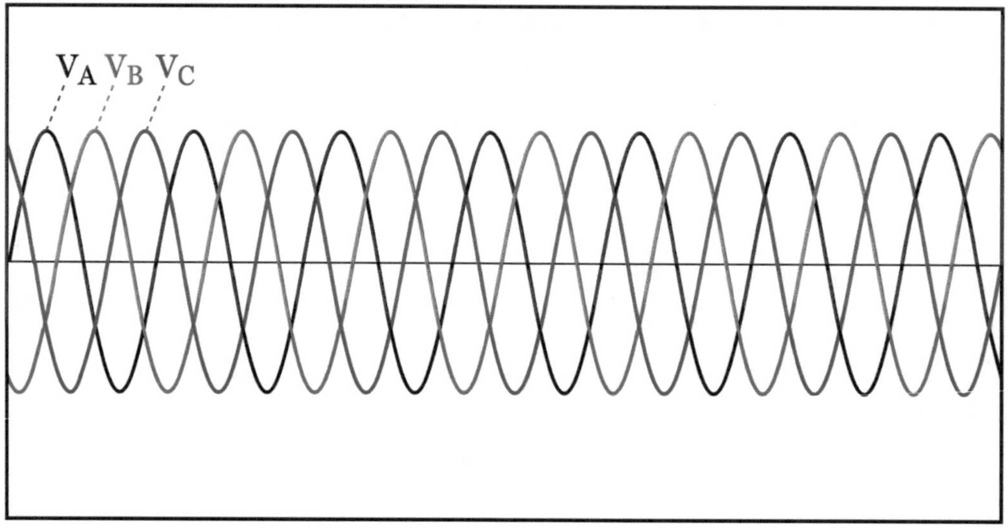

Normal Operating Conditions

If we were looking at the current waveform for normal operating conditions, we would see no dramatic increases in current, only slight changes depending on any loads being switched on or off as part of normal conditions of the electrical system being monitored.

Phase to Phase fault:

The fourth three-phase voltage waveform shows a **phase to phase fault**, also known as a line to line fault or a double line fault.

Notice only the voltage on two phases has decreased at the time of the fault. In this particular example, this is an A-phase to C-phase fault:

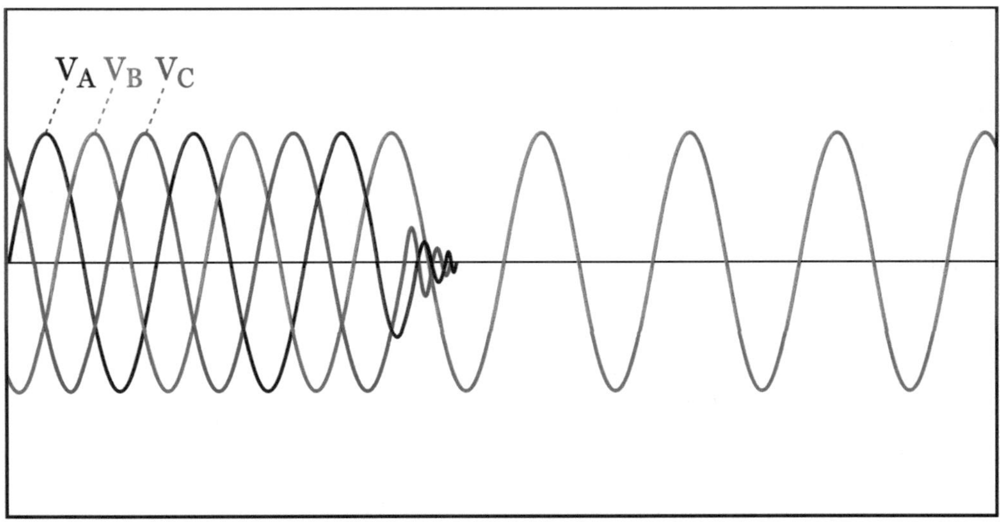

Phase to Phase Fault

If we were looking at the current waveform for this fault, the current on A-phase and C-phase would have dramatically increased at the same time when the A-phase and C-phase voltage decreased.

The answer is: **See completed drag and drop diagram above.**

(Drag and drop AIT question) - Ch. 8.5 Fault Current Analysis

Problem #52 Solution

A delta-wye grounded transformer is a three-phase transformer made from three individual single-phase transformers. The primary terminals of all three single-phase transformers are connected in delta, the secondary terminals of all three single-phase transformers are connected in wye, and finally the neutral terminal of the secondary wye connection is connected directly to ground:

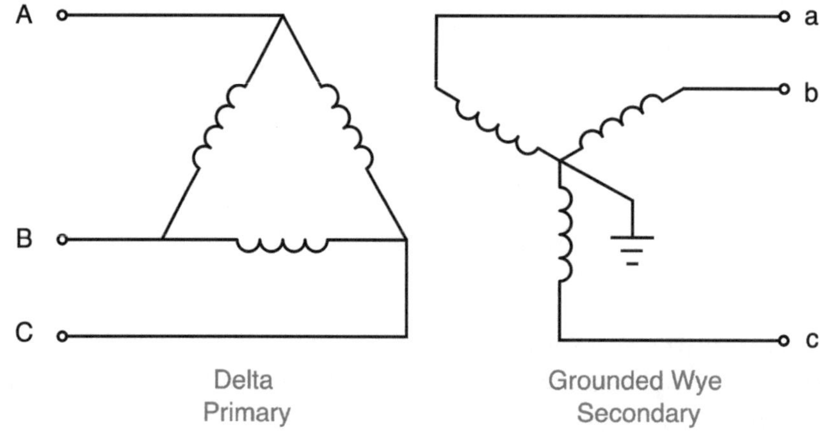

> *Note: We will be using capital letters (A, B, C) for the primary terminals and lowercase letters (a, b, c, n) for the secondary terminals to distinguish between the two sides of the transformer in the following solution.*

Let's evaluate each possible statement to identify which ones apply to a single-phase transmission line.

(A) The secondary line voltage will lead the primary line voltage by 30 degrees.

Let's trace the path of the line voltage on the primary side (V_{AB}), through the transformer to the secondary side (V_{an}):

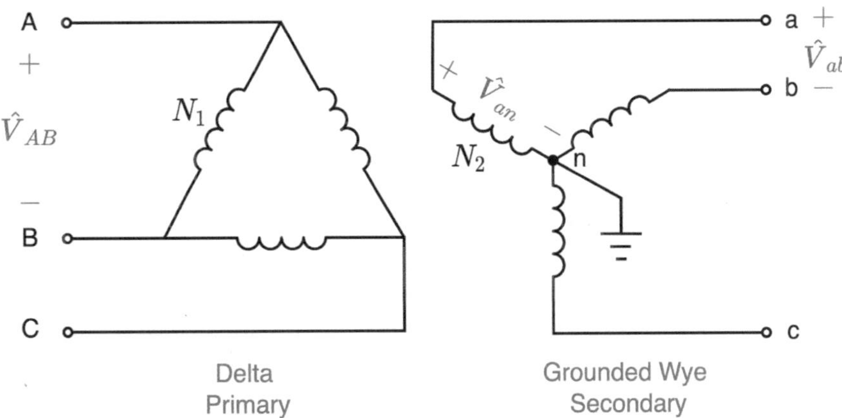

For a **delta** connection, the phase voltage (V_P) directly across each winding of the transformer is equal to the applied line voltage (V_L)[74].

Because of this, the A-phase **delta** primary voltage from terminal A to terminal B is the same as the A line voltage of the system (V_{AB}).

Next, let's step through just one pair of the single-phase windings of the transformer from primary to secondary to solve for the secondary A-phase voltage of the wye connection (V_{an}) from the a terminal to the neutral terminal using the single-phase transformer ratio[75]:

$$\frac{E_1}{E_2} = \frac{N_1}{N_2}$$

$$E_2 = E_1 \frac{N_2}{N_1}$$

$$\hat{V}_{an} = \hat{V}_{AB} \frac{N_2}{N_1}$$

*Note: The primary number of turns (N_1) and the secondary number of turns (N_2) used above represent the turns of each of the three pairs of single-phase transformers that make up the overall three-phase transformer. It does **not** represent the three-phase ratio of the delta-wye transformer.*

We are now on the secondary side of the transformer. For a **wye** connection, the line voltage (V_L) is larger by a factor of the square root of three compared to the wye phase

[74] NCEES® Reference Handbook (Version 1.1.2) - 3.1.1 3-Phase Circuits p. 33
[75] NCEES® Reference Handbook (Version 1.1.2) - 4.3.1.1 Single-Phase Transformer Equivalent Circuits p. 58

voltage, and it leads the wye phase voltage by 30 degrees[76].

We can use this relationship to solve for the secondary line voltage (V_{ab}) from the wye phase voltage (V_{an}):

$$\hat{V}_{ab} = \hat{V}_{an}\left(\sqrt{3} < 30^{\circ}\right)$$

Then, we solve for the secondary line voltage (V_{ab}) from the primary line voltage (V_{AB}) by substituting in the relationship we found earlier for the **wye** phase voltage (V_{an}):

$$\hat{V}_{ab} = \hat{V}_{an}\left(\sqrt{3} < 30^{\circ}\right)$$

$$\hat{V}_{an} = \hat{V}_{AB}\frac{N_2}{N_1}$$

$$\hat{V}_{ab} = \hat{V}_{AB}\frac{N_2}{N_1}\left(\sqrt{3} < 30^{\circ}\right)$$
$$\hat{V}_{ab} = \sqrt{3}|V_{AB}|\frac{N_2}{N_1} < \theta_{V_{AB}} + 30^{\circ}$$

Looking at the final relationship above, notice that the secondary line voltage (V_{ab}) on the **wye** connection will **lead** the primary line voltage of the **delta** connection (V_{AB}) by 30 degrees ($\theta_{VAB} + 30^{o}$).

True! **The secondary line voltage will lead the primary line voltage by 30 degrees.**

(B) The secondary line current will lead the primary line current by 30 degrees.

Let's trace the path of the line current on the primary side (I_A), through the transformer to the secondary side (I_a):

[76] NCEES® Reference Handbook (Version 1.1.2) - 3.1.1 3-Phase Circuits p. 33

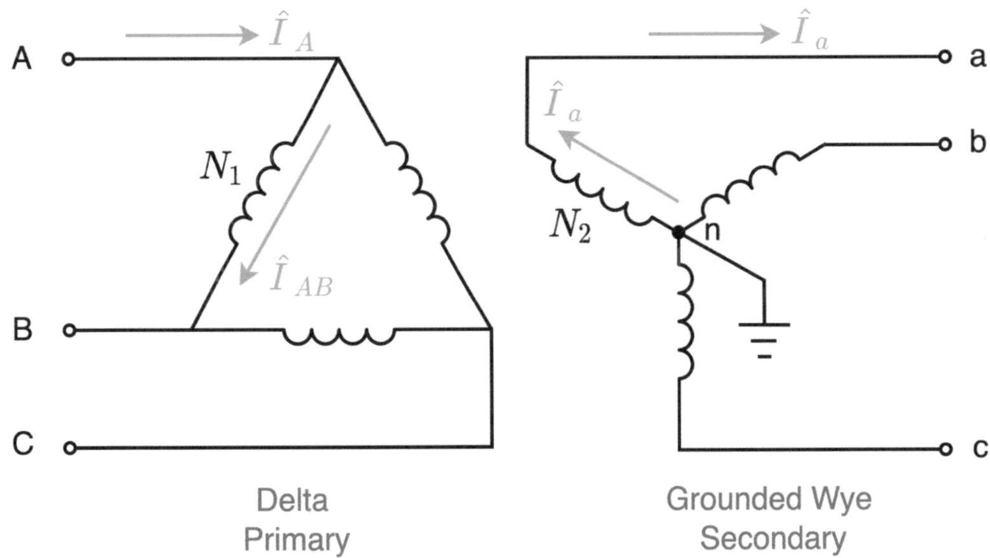

For a **delta** connection, the phase current (I_P) through each winding is smaller by a factor of the square root of three compared to the line current (I_L) entering or leaving the delta connection[77] and leads the line current by 30 degrees.

Let's use this relationship to solve for the primary delta phase current (I_{AB}) from the primary line current entering the delta connection (I_A):

$$\hat{I}_{AB} = \frac{\hat{I}_A}{\sqrt{3}} < 30^\circ$$

Next, let's step through just one pair of the single-phase windings of the transformer from primary to secondary to solve for the secondary phase current of the wye connection (I_a) from the neutral terminal to the a terminal using the single-phase transformer ratio[78]:

$$\frac{I_2}{I_1} = \frac{N_1}{N_2}$$

$$I_2 = I_1 \frac{N_1}{N_2}$$

$$\hat{I}_a = \hat{I}_{AB} \frac{N_1}{N_2}$$

[77] NCEES® Reference Handbook (Version 1.1.2) - 3.1.1 3-Phase Circuits p. 33
[78] NCEES® Reference Handbook (Version 1.1.2) - 4.3.1.1 Single-Phase Transformer Equivalent Circuits p. 58

*Note: The primary number of turns (N_1) and the secondary number of turns (N_2) used above represent the turns of each of the three pairs of single-phase transformers that make up the overall three-phase transformer. It does **not** represent the three-phase ratio of the delta-wye transformer.*

We are now on the secondary side of the transformer. For a **wye** connection, the line current (I_L) is equal to the phase current (I_P)[79].

Let's solve for secondary line current (I_a) leaving the wye connection by substituting in the expression we found earlier for the primary delta phase current (I_{AB}):

$$\hat{I}_a = \hat{I}_{AB}\frac{N_1}{N_2}$$

$$\hat{I}_{AB} = \frac{\hat{I}_A}{\sqrt{3}} < 30^\circ$$

$$\hat{I}_a = \frac{\hat{I}_A}{\sqrt{3}}\frac{N_1}{N_2} < 30^\circ$$

$$\hat{I}_a = \frac{|I_A|}{\sqrt{3}}\frac{N_1}{N_2} < \theta_{I_A} + 30^\circ$$

Looking at the final relationship above, notice that the secondary line current (I_a) leaving the wye connection will **lead** the primary line current (I_A) entering the **delta** connection by 30 degrees ($\theta_{IA} + 30^o$).

True! **The secondary line current will lead the primary line current by 30 degrees.**

(C) There is no phase shift in the secondary line voltage or secondary line current compared to the primary line voltage and primary line current.

While evaluating the answer choices (A) and (B), we learned that both the secondary line voltage and secondary line current will lead their respective primary line voltage and primary line current by 30 degrees.

[79] NCEES® Reference Handbook (Version 1.1.2) - 3.1.1 3-Phase Circuits p. 33

This means that the phase angle for the secondary line voltage and the secondary line current of a delta-wye grounded transformer will be increased by 30 degrees compared to the primary line voltage and the primary line current on the same phase. Note that this is also true for a delta-wye non-grounded transformer.

False! **There IS a phase shift in the secondary line voltage and secondary line current compared to the primary line voltage and primary line current.**

(D) The three-phase delta-wye transformer voltage ratio is smaller by a factor of the square root of three compared to the per-phase voltage ratio of each single-phase transformer inside of the three-phase transformer.

To help answer this, let's use the last expression we arrived at while evaluating answer choice (A) comparing the secondary line voltage (V_{ab}) on the **wye** connection to the primary line voltage of the **delta** connection (V_{AB}):

$$\hat{V}_{ab} = \sqrt{3}|V_{AB}|\frac{N_2}{N_1} < \theta_{V_{AB}} + 30^\circ$$

Since the transformer ratio only effects magnitudes and **not** phase angles, let's simplify the above expression by using magnitudes only instead of complex numbers:

$$|V_{ab}| = \sqrt{3}|V_{AB}|\frac{N_2}{N_1}$$

Next, let's rearrange the expression to solve for the three-phase voltage ratio of the delta-wye grounded transformer:

$$\frac{|V_{AB}|}{|V_{ab}|} = \frac{N_1}{\sqrt{3}N_2}$$

Note: The voltage ratio of a three-phase transformer is the ratio of the primary line voltage to the secondary line voltage ($a_{3\phi} = V_{AB} / V_{ab}$). The per-phase ratio of each individual single-phase transformer that makes up the overall three-phase transformer is the ratio of the primary to secondary phase voltage or the primary to secondary number of turns ($a_{1\phi} = E_1 / E_2 = N_1 / N_2$).

Notice that the three-phase voltage ratio of the delta-wye grounded transformer (V_{AB} / V_{ab}) is equal to the ratio of the individual single-phase transformer winding ratio **divided by the square root of three!**

While the transformer ratings in the problem are a red herring, we can use the given three-phase voltage ratio to further this point for learning purposes.

If the three-phase delta-wye grounded voltage ratio is 4.16k-480V (from the problem statement, 277V is the single-phase voltage on the secondary wye side), then the per-phase voltage ratio of each single-phase transformer inside of the three-phase delta-wye grounded transformer is approximately 7.205k-480V.

We can prove this by starting with the previous expression:

$$\frac{|V_{AB}|}{|V_{ab}|} = \frac{N_1}{\sqrt{3}N_2}$$

Then, substituting in the relationship between the per-phase winding ratio and the single-phase voltage ratio ($N_1/N_2 = E_1/E_2$)[80] followed by plugging in the primary and secondary line voltages and single-phase voltages:

$$\frac{|V_{AB}|}{|V_{ab}|} = \frac{E_1}{\sqrt{3}E_2}$$

$$\frac{4.16\,kV}{480\,V} = \frac{7.205\,kV}{\sqrt{3}(480\,V)}$$

$$8.67 = 8.67$$

The three-phase delta-wye grounded transformer ratio is approximately $a_{3ø} = 8.67$, and the per-phase voltage ratio of each single-phase transformer is approximately 7.205k-480V.

True! **The three-phase delta-wye transformer voltage ratio is smaller by a factor of the square root of three compared to the voltage ratio of the individual single-phase transformers.**

(E) Zero sequence current can flow through the transformer.

Notice that the zero sequence network for a delta-wye grounded transformer[81] has an open circuit on the delta side and a short to ground on the wye grounded side:

---

[80] NCEES® Reference Handbook (Version 1.1.2) - 4.3.1.1 Single-Phase Transformer Equivalent Circuits p. 58
[81] NCEES® Reference Handbook (Version 1.1.2) - 5.1.5 Fault Current Analysis p. 70

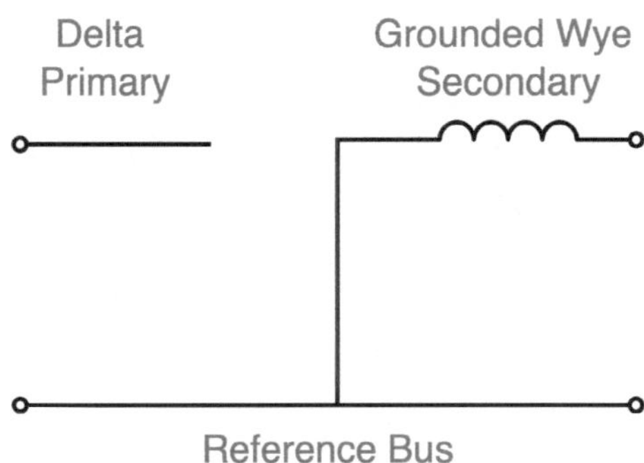

The only transformer connection that zero sequence current can flow through is a wye-wye transformer grounded on both the primary and secondary side (wye-grounded wye-grounded).

False! **Zero sequence current CANNOT flow through the transformer.**

The answer is: **A, B, and D.**

(Multiple correct AIT question) - Ch. 8.6 Transformer Connections

Problem #53 Solution

A single-phase transmission is made up of two equal conductors, the "hot" conductor and the neutral conductor (also known as "the return conductor"). The inductance (L) and capacitance (C) characteristics of a single-phase transmission line can be described using either average values (also known as the "one-way," "per-conductor," or "line-to-neutral" values) or total values.

The average values of a single-phase transmission line represent the characteristics of just one of the conductors, compared to the total values which represent the single-phase transmission line characteristics as a whole.

This is **unlike three-phase transmission lines which can only be described in terms of average, line-to-neutral values.**

Let's evaluate each possible statement to identify which ones apply to a single-phase transmission line.

(A) The average line inductance is twice as large as the total line inductance.

We can model the difference between the average line inductance (L) and the total inductance (L_{Total}) for the single-phase transmission line by drawing a circuit with equal hot and neutral conductors in series with a single-phase load:

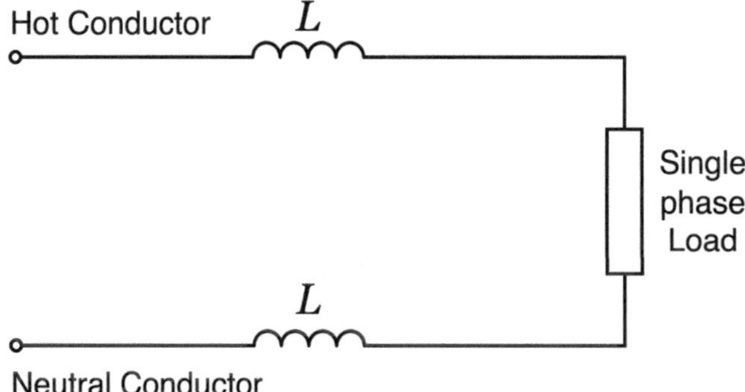

We can simplify the circuit by combining the series inductance of the hot and neutral conductor:

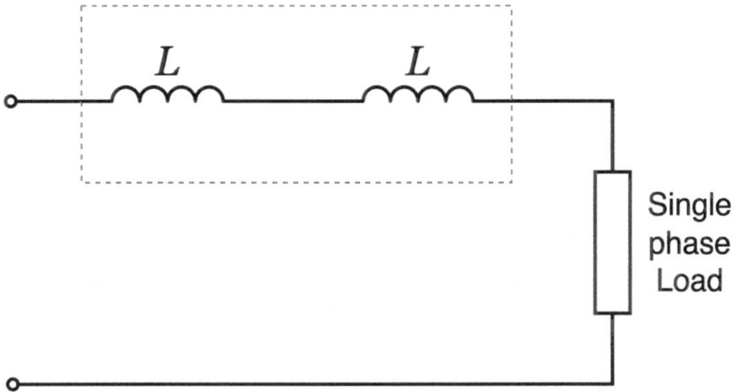

Inductance in series can be combined by summing them together. This will equal the total inductance (L_{Total}) of the single-phase transmission line:

$$L_{Total} = L + L$$
$$L_{Total} = 2L$$

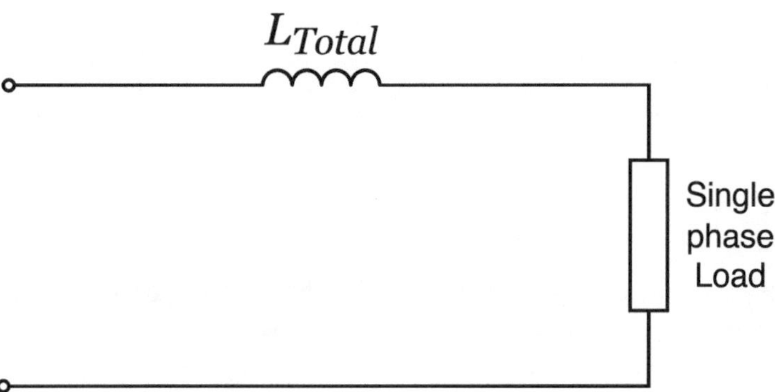

This means that for a single-phase transmission line, the total inductance (L_{Total}) is twice as large as the average inductance (L) and the average inductance (L) is half the total inductance (L_{Total}).

False! **(A) The average line inductance is NOT twice as large as the total line inductance.**

(B) The average line inductive reactance is twice as large as the total line inductive reactance.

We can calculate the average inductive reactance (X_L) of the single-phase transmission line by converting the average inductance (L) from the unit of henries [H] to the unit of ohms [Ω][82]:

$$X_L = wL$$
$$X_L = 2\pi f L$$

Similarly, we can calculate the total inductive reactance ($X_{L\,Total}$) of the single-phase transmission line by converting the the total inductance (L_{Total}) from the unit of henries [H] to the unit of ohms [Ω] using the same formula:

$$X_{L\,Total} = 2\pi f L_{Total}$$

To compare the average inductive reactance (X_L) to the total inductive reactance ($X_{L\,Total}$), we can substitute in the expression for total inductance (L_{Total}) and average inductance (L) that we solved for while evaluating answer choice (A) into the expression above:

$$X_{L\,Total} = 2\pi f L_{Total}$$

$$L_{Total} = 2L$$

$$X_{L\,Total} = 2\pi f(2L)$$
$$X_{L\,Total} = 4\pi f L$$

Notice that for a single-phase transmission line, the total inductive reactance ($X_{L\,Total}$) is twice as large as the average inductive reactance (X_L) and the average inductive reactance (X_L) is half the total inductive reactance ($X_{L\,Total}$).

False! (B) The average line inductive reactance is NOT twice as large as the total line inductive reactance.

(C) The average line capacitance is twice as large as the total line capacitance.

While the properties of inductance (L) are due to the physical material of the conductor, the properties of capacitance (C) are due to two electrically charged conductors in close proximity side by side in parallel.

We can model the shunt capacitance (C) that occurs between the hot and neutral conductors of a single-phase transmission line by drawing a cross section view of the conductors and modeling two capacitors between them, one for each electrically

[82] NCEES® Reference Handbook (Version 1.1.2) - 4.3.1.1 3.1.4.1 AC Circuits - Phasor Transforms of Sinusoids p. 36

charged conductor:

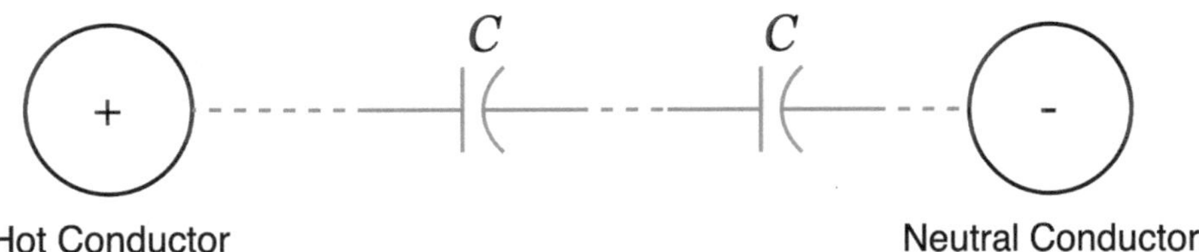

Unlike inductance (L), capacitance (C) in series combines not by summing but by the reciprocal of reciprocal sums. This is the same method as adding resistors (or impedance) in parallel.

We can use this relationship to solve for the total capacitance (C_{Total}) from the average capacitance (C):

$$C_{Total} = \frac{1}{\frac{1}{C} + \frac{1}{C}}$$

$$C_{Total} = \frac{1}{\frac{2}{C}}$$

$$C_{Total} = \frac{1}{2}C$$

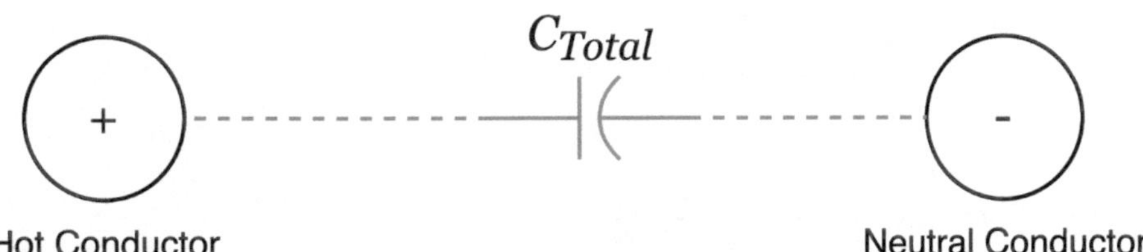

For a single-phase transmission line, the total capacitance (C_{Total}) is half the average capacitance (C) and the average capacitance (C) is twice the total capacitance (C_{Total}).

True! **(C) The average line capacitance is twice as large as the total line**

capacitance.

(D) The average line capacitive reactance is twice as large as the total line capacitive reactance.

We can calculate the average capacitive reactance (X_C) of the single-phase transmission line by converting the average capacitance (C) from the unit of farads [F] to the unit of ohms [Ω][83]:

$$X_C = -\frac{1}{wC}$$

$$X_C = -\frac{1}{2\pi f C}$$

Similarly, we can calculate the total capacitive reactance ($X_{C\,Total}$) of the single-phase transmission line by converting the the total capacitance (C_{Total}) from the unit of farads [F] to the unit of ohms [Ω] using the same formula:

$$X_{C\,Total} = -\frac{1}{2\pi f C_{Total}}$$

To compare the average capacitive reactance (X_C) to the total capacitive reactance ($X_{C\,Total}$), we can substitute in the expression for total capacitance (C_{Total}) and average capacitance (C) that we solved for while evaluating answer choice (C) into the expression above:

$$X_{C\,Total} = -\frac{1}{2\pi f C_{Total}}$$

$$C_{Total} = \frac{1}{2}C$$

$$X_{C\,Total} = -\frac{1}{2\pi f (C/2)}$$

$$X_{C\,Total} = -\frac{1}{\pi f C}$$

83 NCEES® Reference Handbook (Version 1.1.2) - 4.3.1.1 3.1.4.1 AC Circuits - Phasor Transforms of Sinusoids p. 36

Notice that for a single-phase transmission line, the total capacitive reactance ($X_{C\,Total}$) is twice as large as the average capacitive reactance (X_C) and the average capacitive reactance (X_C) is half the total capacitive reactance ($X_{C\,Total}$).

False! **(D) The average line capacitive reactance is NOT twice as large as the total line capacitive reactance.**

(E) Charging current is directly proportional to the total capacitive reactance.

Charging current is the current drawn by a capacitor:

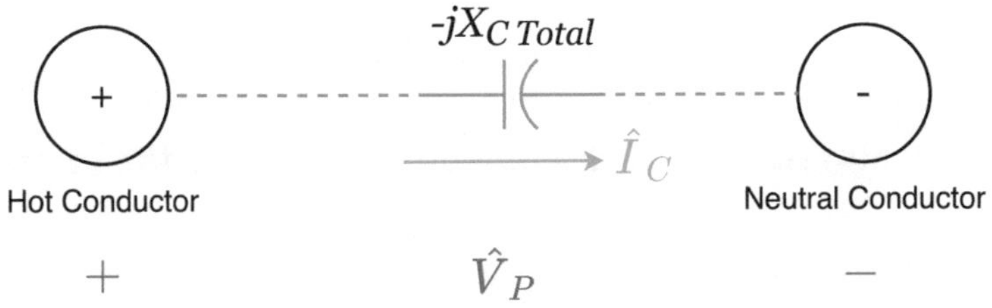

We can solve for the charging current (I_C) using Ohm's law ($V = IZ$) with the total capacitive reactance ($X_{C\,Total}$) between the hot and neutral conductors and the single-phase voltage across the hot and neutral conductors (V_P):

$$\hat{V}_P = -\hat{I}_C \cdot jX_{C\,Total}$$

$$\hat{I}_C = -\frac{\hat{V}_P}{jX_{C\,Total}}$$

$$\hat{I}_C = j\frac{\hat{V}_P}{X_{C\,Total}}$$

Notice that charging current (I_C) is **inversely proportional** (**not** directly proportional) to the total capacitive reactance ($X_{C\,Total}$).

If instead we plug in the expression for the total capacitance (C_{Total}) that we found while evaluating answer choice (D), we can see that this relationship is actually true for capacitance, not reactance:

$$\hat{I}_C = j\frac{\hat{V}_P}{X_{C\,Total}}$$

$$X_{C\,Total} = -\frac{1}{2\pi f C_{Total}}$$

$$\hat{I}_C = j\frac{\hat{V}_P}{-1/(2\pi f C_{total})}$$

$$\hat{I}_C = -j2\pi f V_P C_{Total}$$

An increase or decrease in the total capacitance (C_{Total}) will result in an equal increase or decrease in charging current (I_C) by the same factor.

False! **(E) Charging current is NOT directly proportional to the total capacitive reactance.**

The answer is: **C.**

(Multiple correct AIT question) - Ch. 8.7 Transmission Line Models

Problem #54 Solution

To calculate the real power delivered to customer #2, we will need to first determine the amount of output power (P_{out}) each generator is supplying to the utility bus. Out of the generator ratings given in the single-line diagram, the only value required to calculate the output power (P_{out}) from the given input power (P_{in}) is efficiency (η)[84].

First, let's calculate the output power (P_{out}) of generator #1:

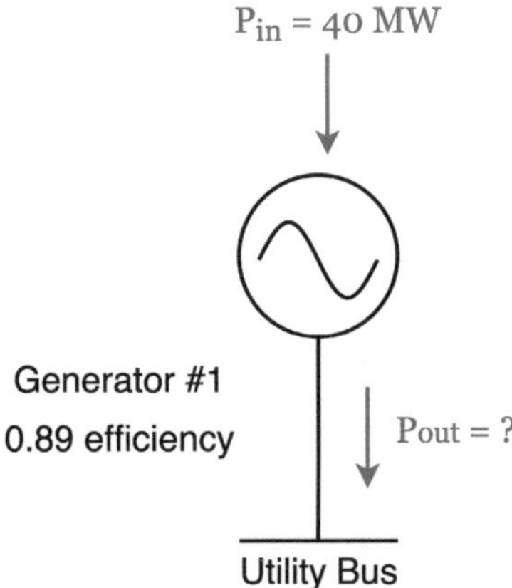

$$\% Efficiency = \frac{P_{out}}{P_{in}} \cdot 100$$

$$P_{out} = \frac{\% Efficiency(P_{in})}{100}$$

$$P_{out} = 0.89 \cdot (40\, MW)$$

```
DEG
0.89*40          35.6
```

Generator #1 supplies 35.6 MW of output power (P_{out}).

Next, let's calculate the output power (P_{out}) of generator #2:

84 *NCEES® Reference Handbook (Version 1.1.2) - 2.2.5 Energy Management p. 27*

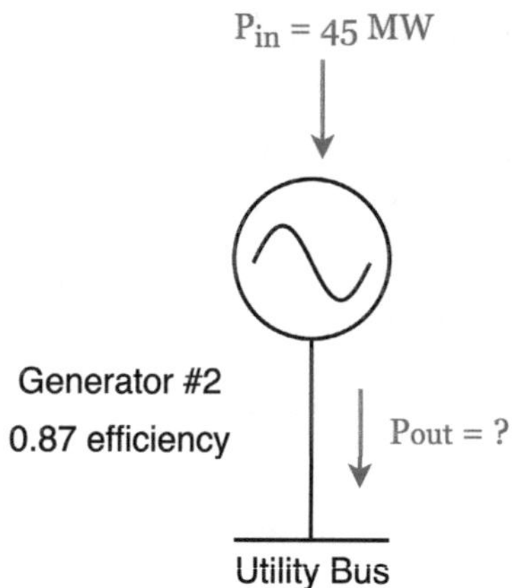

$$\% Efficiency = \frac{P_{out}}{P_{in}} \cdot 100$$

$$P_{out} = \frac{\% Efficiency(P_{in})}{100}$$

$$P_{out} = 0.87 \cdot (45\,MW)$$

DEG
0.87*45 39.15

Generator #2 supplies 39.15 MW of output power (P_{out}).

Let's fill these values into the single-line diagram:

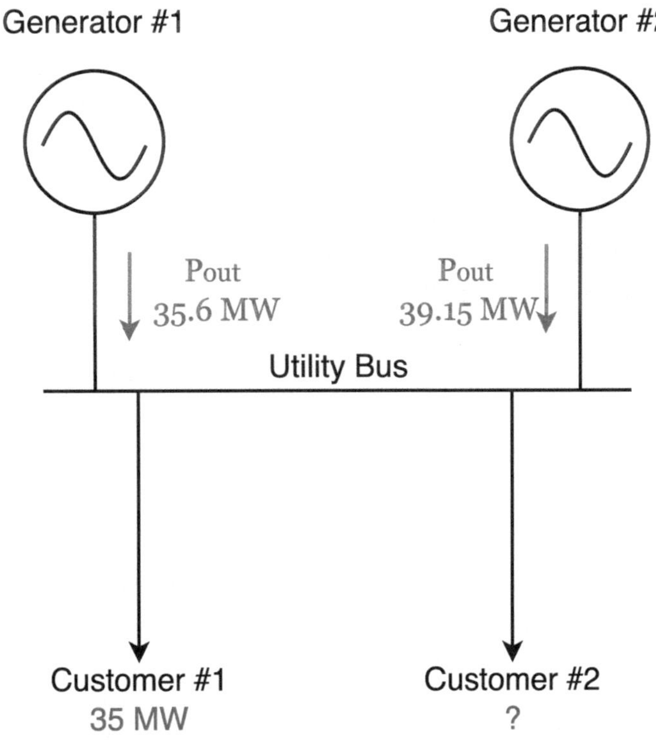

Since we know the amount of real power being supplied by both generators and the amount of real power being consumed by customer #1, we can calculate the remaining power that is being consumed by customer #2:

$$P_{Customer\ \#2} = 35.6\,MW + 39.15\,MW - 35\,MW$$

```
DEG
35.6+39.15-35
39.75
```

The amount of real power consumed by customer #2 rounded to the nearest megawatt is 40 MW.

The answer is: **40.**

(Fill in the blank AIT question) - Ch. 8.8 Power Flow

Problem #55 Solution

The maximum amount of time to clear a fault before a synchronous machine becomes unstable is known as the **critical clearing time (t_{cr})**.

Before we can solve for the critical clearing time (t_{cr}) using the information given in the problem, we'll need to first solve for the **critical clearing angle (δ_{cr})**, which is the maximum electrical torque angle of the machine that the fault must be cleared by in order for it to remain stable.

The critical clearing angle (δ_{cr}) is calculated from the initial rotor torque angle (δ_o)[85] given in the problem:

$$\delta_{cr} = \cos^{-1}\left[(\pi - 2\delta_0)\sin\delta_0 - \cos\delta_0\right]$$

***Careful!* This formula requires using the units of radians!**

Where:

δ_{cr} = critical clearing angle in **radians**.
δ_o = initial rotor torque angle in **radians**.

To use this formula, first change your calculator to radian mode (MODE, RAD):

Next, convert the initial rotor torque angle (δ_o) from degrees to radians:

$$26^\circ \cdot \frac{\pi\,\textbf{rad}}{180^\circ} = 0.454\ \textbf{rad}$$

RAD

$26*\frac{\pi}{180}$ $\frac{13\pi}{90}$

ans

0.453785606

85 *NCEES® Reference Handbook (Version 1.1.2) - 5.1.10 Power System Stability p. 78*

The initial rotor torque angle (δ_o) is 13π/90 radians, or approximately 0.454 radians.

Since this is a fill in your response AIT problem, we need to be as accurate as possible and round only the final value. To use all of the decimals of the initial rotor torque angle (δ_o) in the formula, use the “2nd, answer” feature of the calculator when solving for the critical clearing angle:

$$\delta_{cr} = \cos^{-1}\left[(\pi - 2\delta_0)\sin\delta_0 - \cos\delta_0\right]$$

```
                                      RAD
cos⁻¹((π-2ans)sin(ans)-cos(ans))
                         1.490172515
```

The critical clearing angle (δ_{cr}) is approximately 1.49 radians. Notice the calculator is in radian mode (“RAD” in the upper right hand corner).

We can plot both the initial rotor torque angle (δ_o) and the critical clearing angle (δ_{cr}) on the power-angle curve:

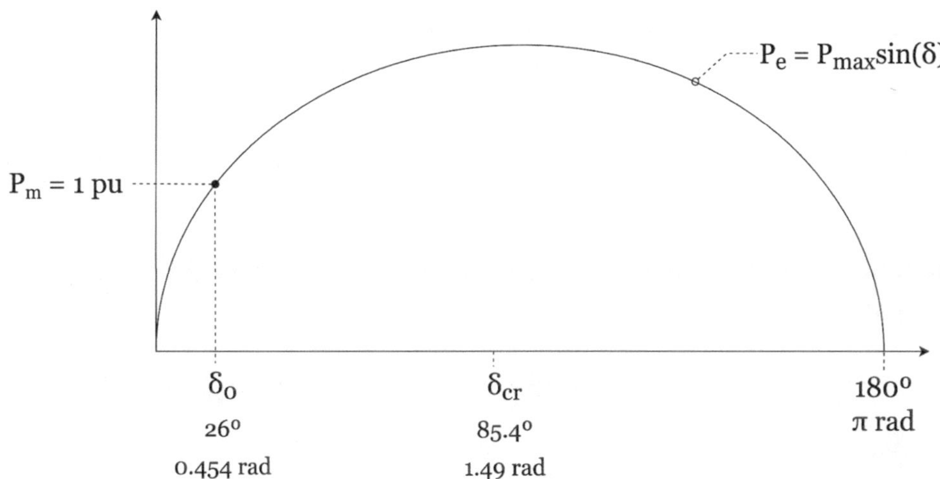

Next, the critical clearing time (t_{cr}) can be calculated from the critical clearing angle (δ_{cr})[86]:

86 *NCEES® Reference Handbook (Version 1.1.2) - 5.1.10 Power System Stability p. 78*

$$t_{cr} = \sqrt{\frac{4H(\delta_{cr} - \delta_0)}{\omega_s P_m}}$$

***Careful!* This formula requires using the units of radians!**

Where:

H = inertia constant in megajoules per megavolt-amperes (MJ/MVA).
δ_{cr} = critical clearing angle in **radians**.
δ_o = initial rotor torque angle in **radians**.
w_s = synchronous speed in radians per second ($w_s = 2\pi f$).
P_m = constant mechanical input power to the machine (watt).

Before we can use this formula, there are two important context clues in the problem:

- At rated conditions, the machine is outputting 100% of its rated power, or P_e = 1 pu.
- At steady state synchronous speed, the mechanical input power (P_m) is equal to the electrical output power (P_e).

Because of this, we will use $P_m = P_{out}$ = 1 pu in the formula above, again using the "2nd, answer" feature of the calculator, this time for the stored critical clearing angle (δ_{cr}):

RAD

$$\sqrt{\frac{4*6\left(ans - 26*\frac{\pi}{180}\right)}{2\pi 60}}$$

0.256862687

The critical clearing time (t_{cr}) is approximately 0.26 seconds rounded to two decimal places.

Note: Don't forget to return your calculator back to degrees (MODE, Deg) before moving on to the next problem:

The answer is: **0.26.**

(Fill in the blank AIT question) - Ch. 8.9 Power System Stability

Problem #56 Solution

Harmonic distortions are usually the result of nonlinear loads, mostly power electronics, that draw current that is not proportional to the supplied voltage at a multiple of the fundamental 60 Hz frequency.

For example, second (2nd) order harmonics are a periodic current at 2 times the fundamental 60 Hz frequency ($f_2 = 2 \times 60\ Hz = 120\ Hz$).

Third (3rd) order harmonics are a periodic current at 3 times the fundamental 60 Hz frequency ($f_3 = 3 \times 60\ Hz = 180\ Hz$).

Nth order harmonics are a periodic current at n times the fundamental 60 Hz frequency ($f_n = n \times 60\ Hz$).

Let's evaluate each possible statement to identify which ones apply to impedance and changes in frequency due to harmonics.

(A) Inductive reactance is directly proportional to frequency.

Inductive reactance (X_L) is determined by inductance (L) and frequency (f)[87]:

$$X_L = wL$$

$$X_L = 2\pi f L$$

Notice that any changes in frequency (f) will result in an equal change in inductive reactance (X_L) by the same factor. For example, if frequency (f) were to increase by a factor of 2, inductive reactance (X_L) would **increase** by a factor of 2. This is known as being **directly proportional**.

True! **(A) Inductive reactance is directly proportional to frequency.**

(B) Capacitive reactance is directly proportional to frequency.

Capacitive reactance (X_C) is determined by capacitance (C) and frequency (f)[88]:

[87] *NCEES® Reference Handbook (Version 1.1.2) - 3.1.4.1 AC Circuits - Phasor Transforms of Sinusoids p. 36*
[88] *NCEES® Reference Handbook (Version 1.1.2) - 3.1.4.1 AC Circuits - Phasor Transforms of Sinusoids p. 36*

$$X_C = -\frac{1}{wC}$$

$$X_C = -\frac{1}{2\pi fC}$$

Notice that any changes in frequency (f) will result in an equal but **opposite** change in capacitive reactance (X_C) by the same factor. For example, if frequency (f) were to increase by a factor of 2, capacitive reactance (X_C) would **decrease** by a factor of 2. This is known as being **inversely proportional.**

False! **(B) Capacitive reactance is NOT directly proportional to frequency.**

(C) Resistance is directly proportional to frequency.

Unlike inductive reactance (X_L) and capacitive reactance (X_C), resistance (R) does not change due to frequency (f). The resistance at any frequency is the same.

False! **(C) Resistance is NOT directly proportional to frequency.**

(D) Impedance is directly proportional to frequency.

Impedance (Z) is the sum of resistance (R) and reactance (X):

$$\hat{Z} = R \pm jX$$

Although reactance (X) is directly proportional with frequency (f) if it is inductive, and inversely proportional with frequency (f) if it is capacitive, resistance does not change with frequency (f).

This means that impedance (Z) overall is NOT directly proportional with frequency (f), even if the reactance (X) is inductive.

False! **(D) Impedance is NOT directly proportional to frequency.**

(E) Reactance at the third harmonic will be three times as large compared to the fundamental frequency if it is inductive.

We already learned while evaluating answer choice (A) that inductive reactance is directly proportional to frequency. This means that the third harmonic ($n = 3$) would result in inductive reactance increasing three times in value (becoming three times as large).

Let's start with the formula for inductive reactance (X_L):

$$X_L = 2\pi f L$$

The 3rd harmonic frequency (f_3) is three times as large ($n = 3$) as the fundamental frequency (f):

$$f_3 = 3f$$

Substituting this into the formula for inductive reactance (X_L) results in a reactance that is three times as large:

$$X_L = 2\pi f L$$

$$f_3 = 3f$$

$$X_{L\,f3} = 2\pi(3f)L$$

$$X_{L\,f3} = 6\pi f L$$

For example, let's calculate the reactance for a 5Ω inductive reactance at the third harmonic by first calculating the inductance (L) in henries [H]:

$$L = \frac{X_L}{2\pi f}$$

$$L = \frac{5\,\Omega}{2\pi \cdot 60\,Hz}$$

$$L = 0.0133\,H$$

Next, calculate the new inductive reactance (X_L) at the third harmonic frequency (f_3) of 180 Hz (3 × 60 Hz = 180 Hz):

$$X_L = 2\pi f L$$

$$X_L = 2\pi(180\,\text{Hz})(0.0133\,H)$$

$$X_L = 15\,\Omega$$

The inductive reactance (X_L) has increased to 15 ohms at 180 hertz, compared to 5 ohms at 60 hertz, which is an increase by a factor of 3 (5Ω × 3 = 15Ω).

True! **(E) Reactance at the third harmonic will be three times as large compared to the fundamental frequency if it is inductive.**

The answer is: **A and E.**

(Multiple correct AIT question) - Ch. 8.4 Power Quality

Problem #57 Solution

Surge impedance loading (SIL) of a transmission line is the amount of power the transmission line can deliver when the resistance (R) and conductance (B) of the line is neglected and the load on the transmission line is equal to surge impedance (Z_s).

Just as resistance (R) is the real component of impedance (Z), conductance (B) is the real component of admittance (Y). When resistance (R) and conductance (B) are neglected, the line is considered to be lossless since the real power losses in watts are being ignored. This is helpful when analyzing a transmission line for high-frequency conditions or for lightning strike surges.

The surge impedance loading (SIL) can be calculated from the magnitude of the line voltage (V_L) at the load and the magnitude of the surge impedance (Z_S)[89]:

$$SIL = \frac{V_{\text{line}}^2}{Z_s}$$

Surge impedance (Z_S) is the characteristic impedance (Z_c) of a lossless line and is calculated using the same formula[90]:

$$Z_s = Z_c = \sqrt{\frac{z}{y}}$$

Let's calculate the surge impedance (Z_S) using the series impedance (z = 30 + j135 ohms) and shunt admittance (y = j750 microsiemens) given in the problem:

$$Z_s = \sqrt{\frac{j135\,\Omega}{j750\,uS}}$$

> ***Careful!*** *Be sure to neglect the resistance (R) of the line impedance or the conductance (B) of the line admittance (Y) if given when calculating the surge impedance (Z_S) in order to assume lossless conditions, otherwise you'll accidentally calculate the characteristic impedance (Z_c) of the line instead.*

[89] *NCEES® Reference Handbook (Version 1.1.2) - 5.1.6 Transmission Line Models p. 72*
[90] *NCEES® Reference Handbook (Version 1.1.2) - 5.1.6 Transmission Line Models p. 71*

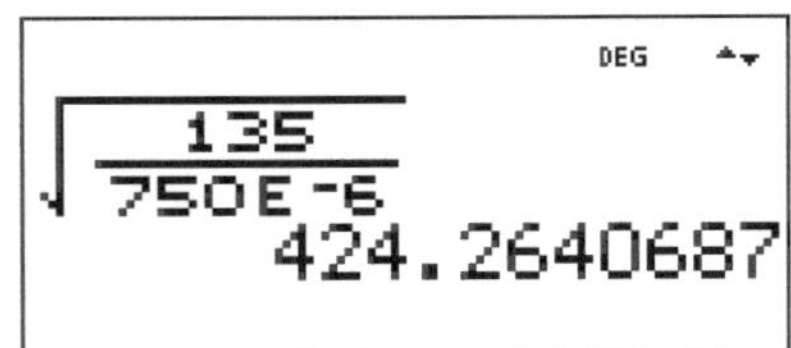

The surge impedance (Z_S) is approximately 424.26 ohms.

Next, let's calculate the surge impedance loading (SIL) using the given line voltage at the receiving load (V_L = 220 kV):

$$SIL = \frac{V_{\text{line}}^2}{Z_s}$$

$$SIL = \frac{220\,kV^2}{424.26\,\Omega}$$

Use the "2nd, answer" feature of your calculator to be as accurate as possible since it is a fill in the blank AIT question:

DEG
220E3²
ans
114079894

The surge impedance loading (SIL) of the transmission line rounded to the nearest megawatt is 114 MW.

The answer is: **114.**

(Fill in the blank AIT question) - Ch. 8.7 Transmission Line Models

Problem #58 Solution

Overcurrent protection is the most basic and fundamental type of protection that is designed to safely interrupt an electrical circuit to protect equipment and people.

The official definition of overcurrent from the *NEC®* is:

> *"NEC® 100 - Overcurrent: Any current in excess of the rated current of equipment or the ampacity of a conductor. It may result from overload, short circuit, or ground fault."*

Let's evaluate each possible statement to identify which types of occurrences overcurrent protection is intended to interrupt.

(A) Short circuit

A short circuit is a sudden unintended electrical connection with a return path back to the power supply. Short circuits typically result in very large current (fault current) due to what is usually a low impedance return path.

Short circuit currents, or fault currents, are very large compared to the rated equipment current or load current.

True! **(A) Short circuit is an occurrence that overcurrent protection is intended to interrupt.**

(B) Ground fault

A ground fault occurs anytime an energized part of an electrical system comes into contact with a grounded connection resulting in a short circuit to ground. A ground fault is a specific type of short circuit to ground.

Depending on the type of system grounding, ground faults can be very large and will typically result in more than the rated current on the grounded phase.

The official definition of ground fault from the *NEC®* is:

"NEC® 100 - Ground Fault. An unintentional, electrically conductive connection between an ungrounded conductor of an electrical circuit and the normally non-current-carrying conductors, metallic enclosures, metallic raceways, metallic equipment, or earth."

True! **(B) Ground fault is an occurrence that overcurrent protection is intended to interrupt.**

(C) Overload

Compared to short circuits, overloads are more gradual in nature and are typically the result of overheating, lack of ventilation, crowding of equipment, or aging equipment.

While overload current is usually nowhere near as excessive as a short circuit, it is still in excess of the equipment rated current and, if not interrupted, can lead to damage.

"NEC® 100 - Overload. Operation of equipment in excess of normal, full load rating, or of a conductor in excess of rated ampacity that, when it persists for a sufficient length of time, would cause damage or dangerous overheating. A fault, such as a short circuit or ground fault, is not an overload."

True! **(C) Overload is an occurrence that overcurrent protection is intended to interrupt.**

(D) Under frequency

Under- and over-frequency occurrences are usually the result of generator speed control issues either due to the speed control governor malfunctioning or sudden changes in load conditions.

While under- and over-frequency conditions may result in overcurrents if not interrupted, overcurrent protection is not intended to interrupt frequency conditions. Instead, this is the job for ANSI #81O - Over Frequency and ANSI #81U - Under Frequency relays.

False! **(D) Under frequency is NOT an occurrence that overcurrent protection is intended to interrupt.**

(E) Voltage transients

Transient voltages are very large in magnitude and short in duration, and, if not interrupted, they can cause damage to equipment and pose personnel safety risk. The most common cause of transient voltage is lightning strike; however, arcing and large reactive loads, such as generators, motors, and capacitors, switching on or off can also be the cause.

While transient voltages may result in overcurrents if not interrupted, overcurrent protection is **not** intended to interrupt this condition. Instead, surge protective devices (SPDs) and surge arrestors (SA) are designed to safely shunt voltage transients to ground, effectively bypassing the electrical system.

False! **(E) Voltage transients are NOT occurrences that overcurrent protection is intended to interrupt.**

The answer is: **A, B and C.**

(Multiple correct AIT question) - Ch. 9.1 Overcurrent Protection

Problem #59 Solution

The *NCEES® Reference Handbook* has a small number of ANSI device numbers[91]. You'll need to be familiar with how each of the device numbers function in order to properly answer this question.

Completed drag and drop diagram:

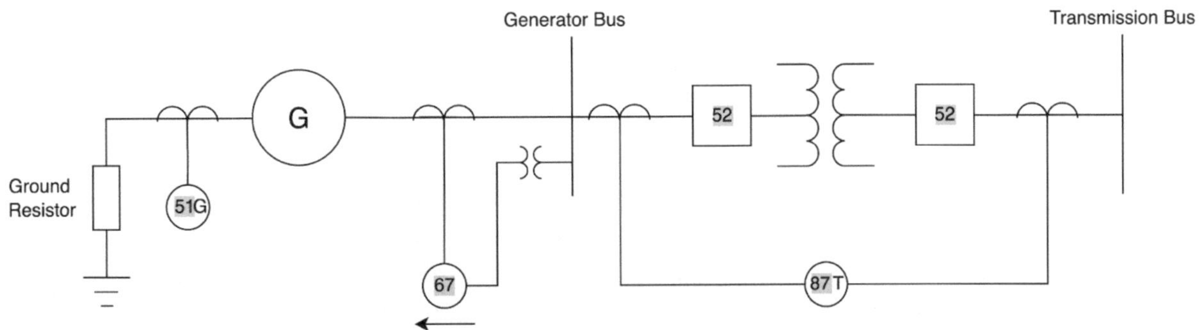

From left to right:

ANSI #51 Inverse Time Overcurrent Relay

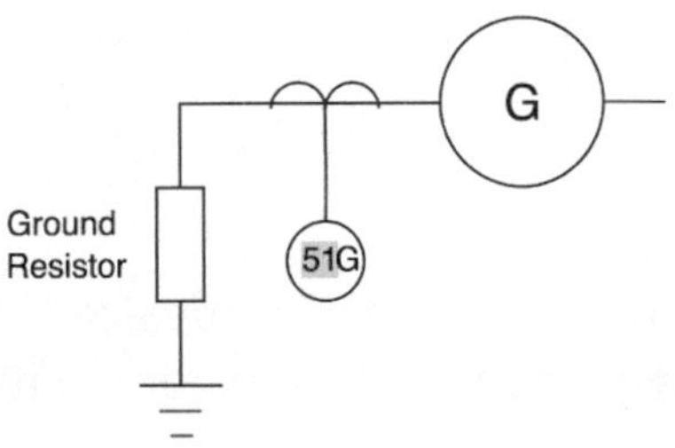

This is an overcurrent relay with an inverse time curve. ANSI #51 and ANSI #50 instantaneous overcurrent relays are the two most common and widely used relay functions. Ground protection when connected directly to the system neutral like the generator in the diagram above, is typically going to be either an ANSI #51 relay, an ANSI #50 relay, or both an ANSI #51 relay and an ANSI #50 relay.

The ANSI "G" suffix can be used for either ground or generator; in this case, it is being used for both.

[91] *NCEES® Reference Handbook (Version 1.1.2) - 3.1.6 Single-Line Diagrams p. 38*

ANSI #67 AC Directional Overcurrent Relay

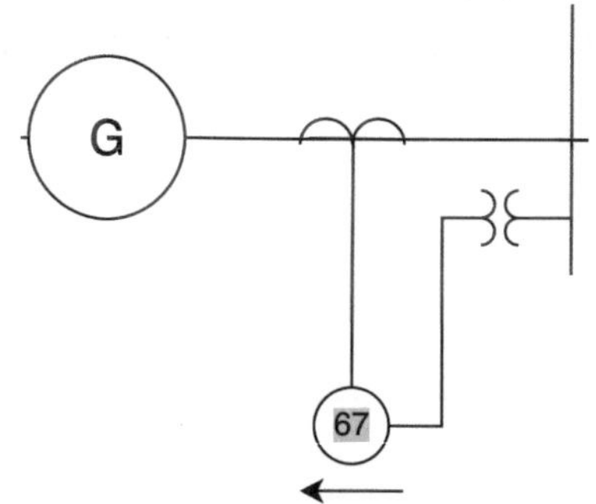

This is an overcurrent relay for fault current flowing in a specific trip direction only. Directional overcurrent relays require both a current input from a current transformer (CT) and a voltage input from a potential transformer (PT). In this example, the ANSI #67 relay is looking for fault current flowing back into the generator indicated by the direction of the arrow next to the relay. Directional overcurrent relays **will not trip** for current flowing in the normal non-tripping direction.

ANSI #52 Circuit Breaker

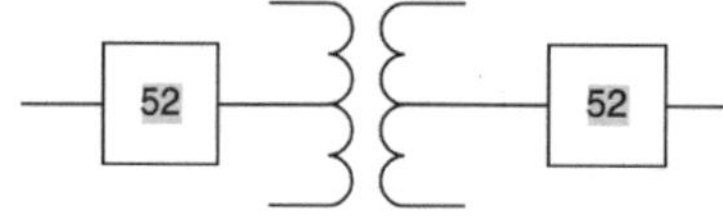

This is a power circuit breaker, the protective device that opens and closes the circuit. ANSI #52 circuit breakers are operated by protective relays through a trip signal.

ANSI #87 Differential Relay

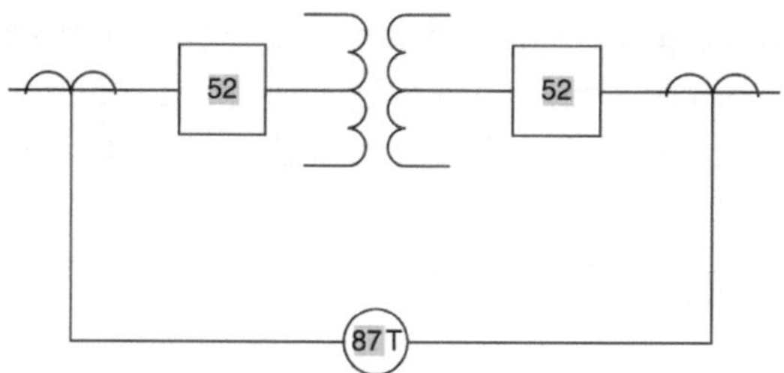

Differential relays monitor current from two different sets of current transformers in order to identify fault current inside the differential zone located between both sets of current transformers.

The ANSI “T” suffix is used for transformer.

The answer is: **See completed drag and drop diagram above.**

(Drag and drop AIT question) - Ch. 9.2 Protective Relaying

Problem #60 Solution

The time current curve (TCC) in the problem is for ground fault overcurrent protection. Answering this problem correctly requires being familiar with TCC basics, logarithmic axis scales, and converting time to cycles:

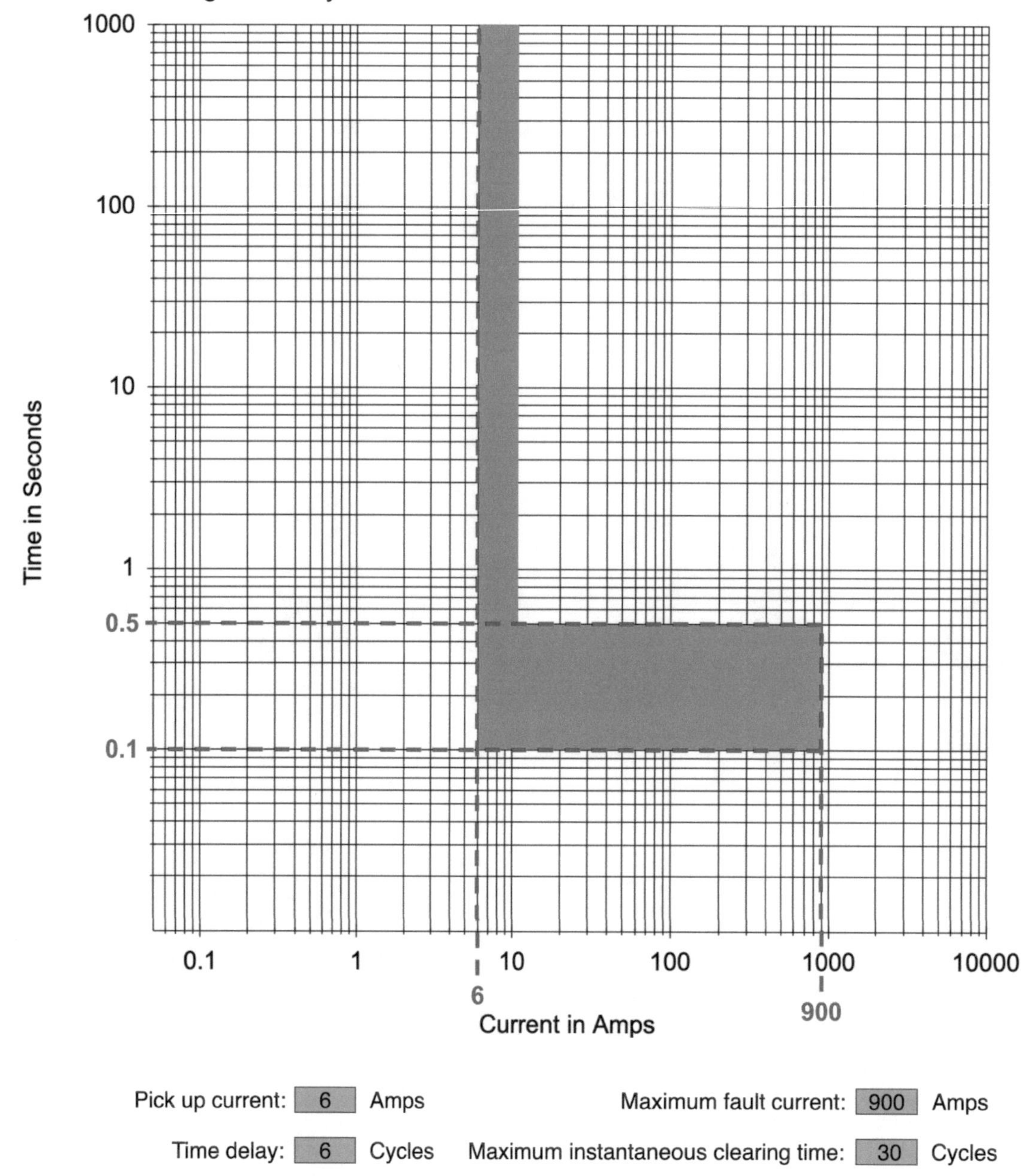

Pick Up Current:

The **pick up current** is the **inside vertical** boundary on the TCC from left to right. This is the amount of current required for the relay to send a trip signal to the device to begin unlatching in order to clear a fault.

Since the horizontal axis is in logarithmic scale, it will be up to you to count up between 1 amp and 10 amps since minor vertical gridlines are not labeled on the TCC graph, which is standard:

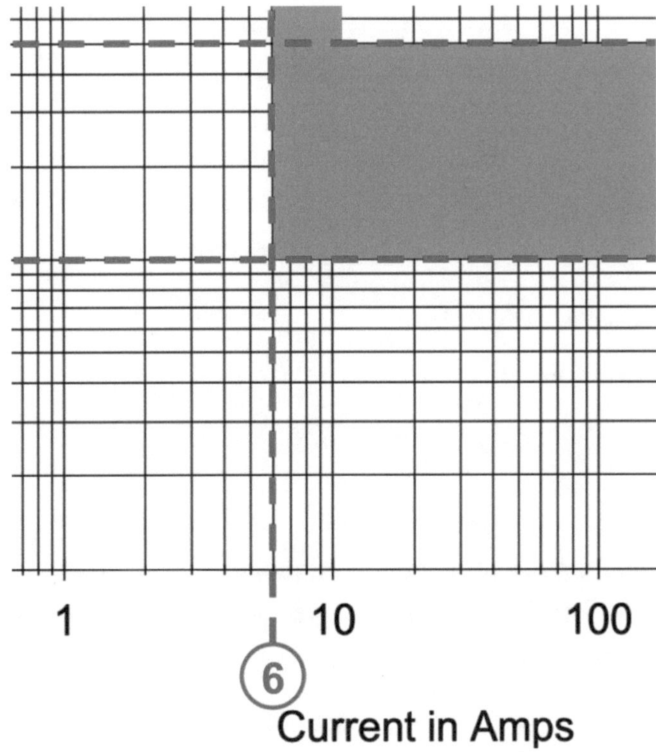

The **pick up current** for this device is **6 amps**, shown circled in **red** above.

Maximum Fault Current:

The **maximum fault current** that the device can interrupt is the **outside vertical** boundary on the TCC from left to right.

This time we must count the minor unlabeled vertical gridlines between 100 amps and 1,000 amps:

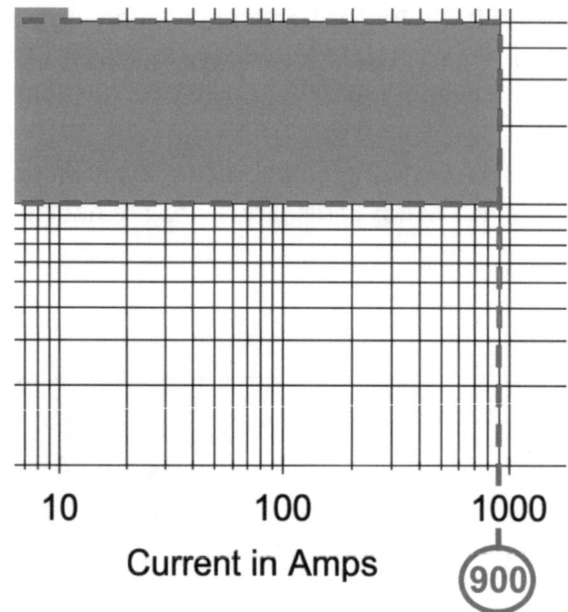

The **maximum fault current** for this device is **900 amps**, shown circled in red above.

Time Delay:

The **time delay** is the intentional amount of time that the relay will wait while it picks up the fault current before sending a trip signal. It is the **inside horizontal** boundary on the TCC from bottom to top.

Just like the horizontal axis, the vertical axis is also in logarithmic scale and the minor horizontal gridlines are not labeled. However, since the time delay falls on a major gridline that is labeled, it is clear to see that **the time delay** is set for **0.1 seconds**, shown circled in red below:

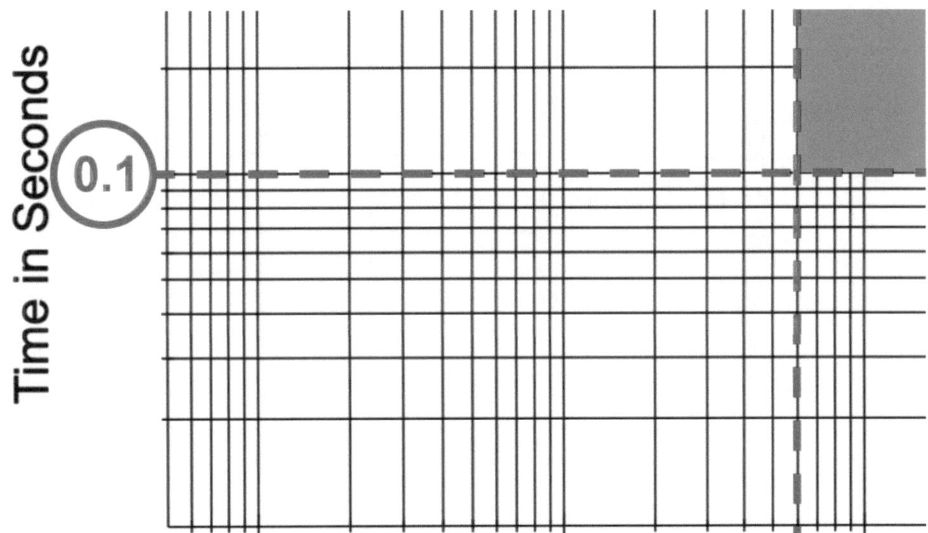

Since the problem asks for time delay in **cycles** (instead of seconds), we need to convert. For a 60 hertz system, which is standard in the United States, there are 60 cycles per second. Since the unit of hertz is the inverse of seconds, it is very easy to convert between the unit of frequency and time:

$$60 \text{ cycles} = 1 \ \mathbf{sec}$$

$$0.1 \ \mathbf{sec} \cdot \frac{60 \text{ cycles}}{1 \ \mathbf{sec}} = 6 \, \text{cycles}$$

The **time delay** is **6 cycles.**

Maximum Instantaneous Clearing Time:

The **maximum instantaneous clearing time** is the maximum amount of time that the device will take to clear a fault in the **instantaneous trip region** after the fault has occurred. It is the **outside horizontal** boundary on the TCC from bottom to top.

Since the maximum instantaneous clearing time is on a minor unlabeled horizontal gridline, we must count on the logarithmic scale between the labeled horizontal gridlines of 0.1 seconds and 1 second:

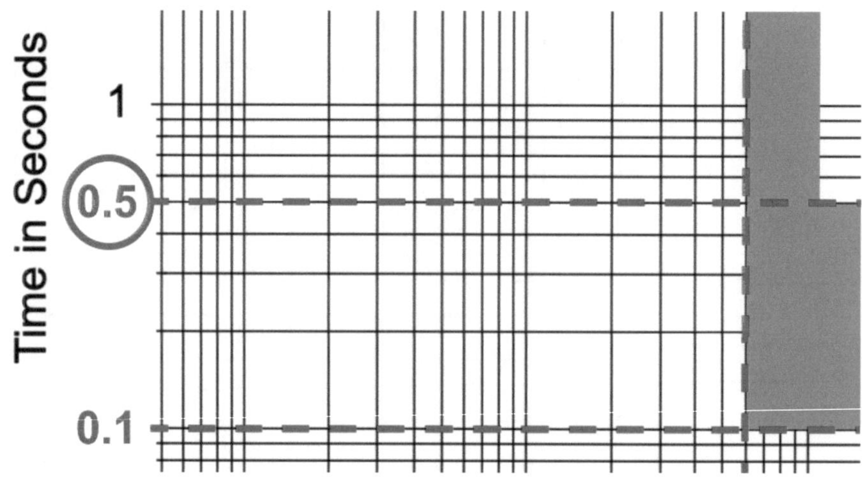

The **maximum instantaneous clearing time** is **0.5 seconds**, shown above circled in red. Again, since the problem asks for cycles instead of seconds, we must convert units:

$$0.5 \textbf{ sec} \cdot \frac{60 \text{ cycles}}{1 \textbf{ sec}} = 30 \, \text{cycles}$$

The **maximum instantaneous clearing time** is **30 cycles.**

The completed drag and drop diagram once more is:

Pick up current: 6 Amps   Maximum fault current: 900 Amps

Time delay: 6 Cycles   Maximum instantaneous clearing time: 30 Cycles

The answer is: **See completed drag and drop diagram above.**

(Drag and drop AIT question) - Ch. 9.1 Overcurrent Protection

Problem #61 Solution

There are many different ways to measure ground fault current. Among them is **core balance ground fault protection**, which is best suited to applications that require very sensitive ground fault protection.

Ground fault applications typically use **ANSI #50** instantaneous overcurrent, **ANSI #51** time delay overcurrent, or a combination of both[92].

Let's evaluate each possible statement to identify which apply to core balance ground fault protection.

(A) One current transformer is used for all current carrying conductors

Below is a diagram for core balance ground fault protection:

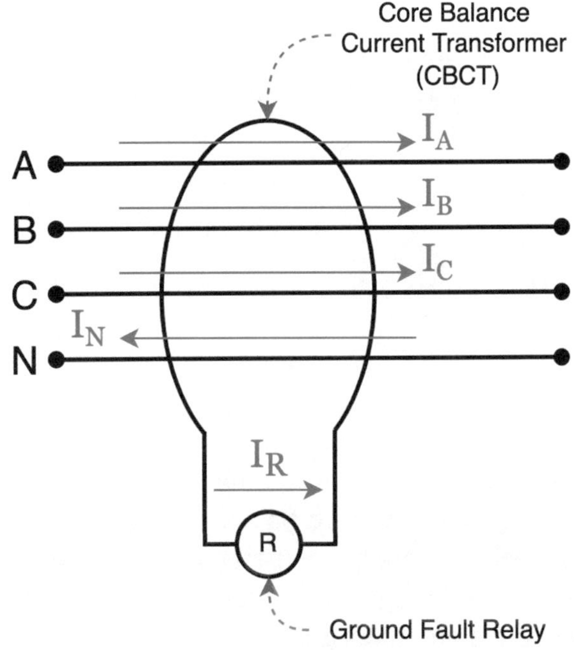

Every current carrying conductor in the circuit is passed through a special type of current transformer known as a **core balance current transformer (CBCT)**, also commonly called a "window current transformer" or a "zero sequence transformer."

True! **(A) One current transformer is used for all current carrying conductors with**

[92] *NCEES® Reference Handbook (Version 1.1.2) - 3.1.6 Single-Line Diagrams p. 38*

core balance ground fault protection.

(B) Measures zero sequence current

Like a normal current transformer, the flux of each conductor induces a secondary current in the relay circuit. Core balance ground fault protection works by summing together the induced secondary current of every conductor that passes through it, **the sum of which is proportional to the zero sequence current in the circuit.**

From symmetrical components, the amount of zero sequence current (I_o) on each phase is identical and equal to one third of the sum of all phase currents[93]:

$$\hat{I}_0 = \hat{I}_{A0} = \hat{I}_{B0} = \hat{I}_{C0}$$

$$\hat{I}_0 = \frac{1}{3}\left(\hat{I}_A + \hat{I}_B + \hat{I}_C\right)$$

This also means that the neutral current (I_N) is equal to three times the zero sequence current (I_o) since neutral current (I_N) is the sum of the phase currents:

$$\hat{I}_N = \hat{I}_A + \hat{I}_B + \hat{I}_C$$

$$\hat{I}_0 = \frac{1}{3}\left(\hat{I}_N\right)$$

$$\hat{I}_N = 3\hat{I}_0$$

During normal conditions or when faults other than a ground fault are present, all current in both directions flows through the core balance current transformer, which results in a net flux of zero. Because of this, the neutral current (I_N) will always equal the sum of the phase currents (this is true even for unbalanced loads) and the current seen by the relay (I_R) will be zero since the two terms are equal and cancel:

$$\hat{I}_R = \hat{I}_A + \hat{I}_B + \hat{I}_C - \hat{I}_N$$

$$\hat{I}_R = 3\hat{I}_0 - \hat{I}_N$$

$$\hat{I}_R = 0$$

---

93 *NCEES® Reference Handbook (Version 1.1.2) - 3.1.2 Symmetrical Components) p. 34*

When a ground fault is present, the current on the faulted phase no longer returns on the neutral (N) and instead bypasses it to ground:

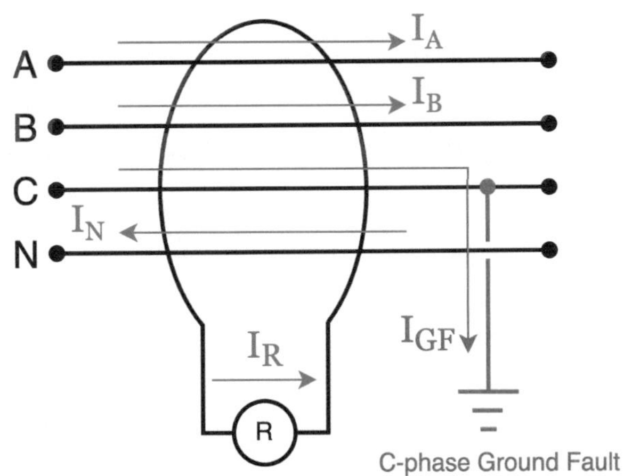

Because the faulted phase current now bypasses the neutral connection at the load, it does not return on the neutral conductor (N) through the core balance transformer, resulting in a current (I_N) that is now smaller than three times the zero sequence current (I_o).

When this happens, the current seen by the relay (I_R) is now greater than zero:

$$\hat{I}_R = 3\hat{I}_0 - \hat{I}_N$$

$$\hat{I}_N < 3\hat{I}_0$$

$$\hat{I}_R > 0$$

The relay (R) will now send a trip signal to the connected circuit breaker to interrupt the circuit due to ground fault.

True! **(B) Core balance ground fault protection measures zero sequence current.**

(C) Measures negative sequence current

False! **(C) Core balance ground fault protection measures zero sequence current, NOT negative sequence current.**

(D) Measures residual current

Residual ground fault protection is a seperate method that measures ground fault current by summing the total current in the neutral of a wye connected current transformer:

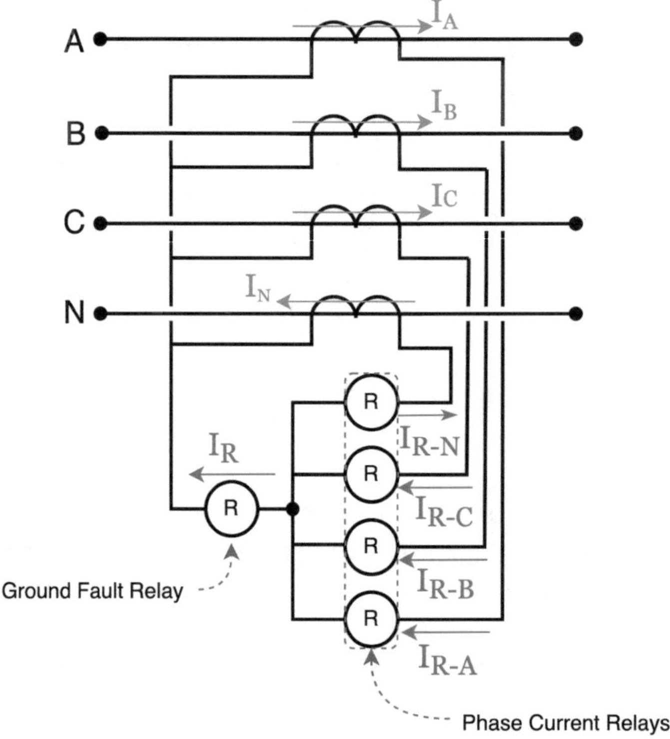

The term **residual** is used to refer to the neutral connection of three wye connected current transformers to differentiate from the neutral connection of the system.

This is similar to **core balance ground fault protection**, and the math is essentially the same. The current seen by the relay (I_R) leaving the **residual connection** is the sum of the relay phase currents ($I_{R\text{-}A}$, $I_{R\text{-}B}$, $I_{R\text{-}C}$) and the relay neutral current ($I_{R\text{-}N}$) according to Kirchoff's current law:

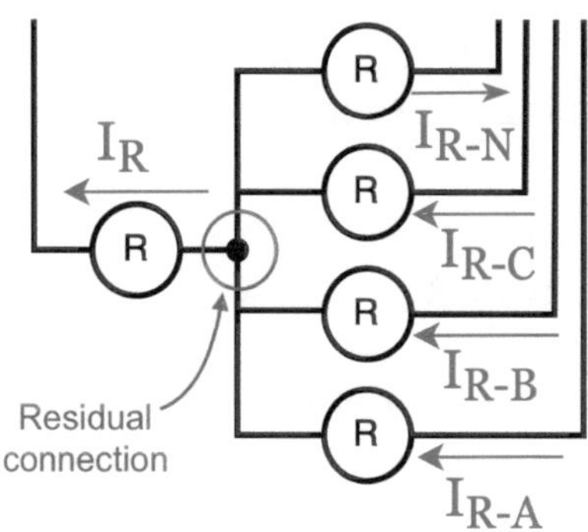

$$\hat{I}_R = \hat{I}_{R-A} + \hat{I}_{R-B} + \hat{I}_{R-C} - \hat{I}_{R-N}$$

During normal conditions or when faults other than a ground fault are present, the sum of the relay phase currents ($I_{R\text{-}A}$, $I_{R\text{-}B}$, $I_{R\text{-}C}$) will equal the relay neutral current ($I_{R\text{-}N}$) and the current seen by the relay (I_R) will equal zero.

When a ground fault is present, the current on the faulted phase no longer returns on the neutral (N) and instead bypasses it to ground:

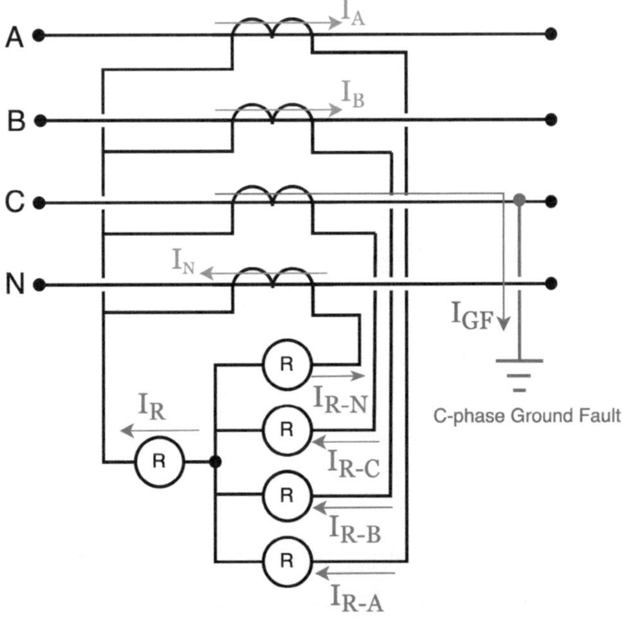

Because the faulted phase current now bypasses the neutral connection at the load, it does not return on the neutral conductor (N), resulting in a relay neutral current ($I_{R\text{-}N}$) that is now smaller than the sum of the relay phase currents ($I_{R\text{-}A}$, $I_{R\text{-}B}$, $I_{R\text{-}C}$).

When this happens, the current seen by the relay (I_R) is now greater than zero:

$$\hat{I}_R = \hat{I}_{R-A} + \hat{I}_{R-B} + \hat{I}_{R-C} - \hat{I}_{R-N}$$

$$\hat{I}_{R-N} < \left(\hat{I}_{R-A} + \hat{I}_{R-B} + \hat{I}_{R-C}\right)$$

$$\hat{I}_R > 0$$

The relay (R) will now send a trip signal to the connected circuit breaker to interrupt the circuit due to ground fault.

False! **(D) Core balance ground fault protection does NOT measure residual current, residual ground fault protection does.**

(E) Connects directly to the system grounded neutral

Ground return ground fault protection is a seperate method that measures ground fault current by connecting directly to the system ground at the service transformer:

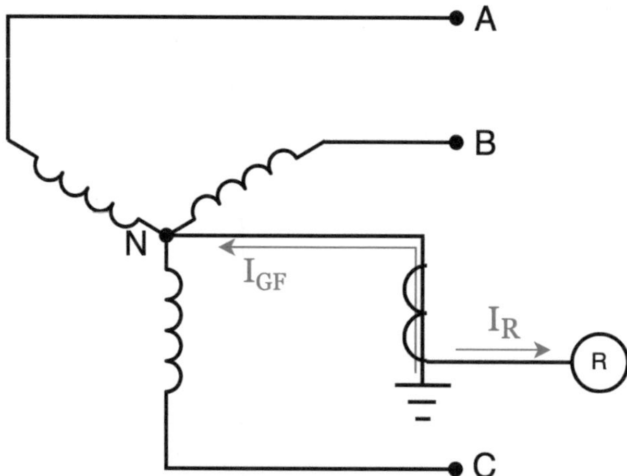

Any current flowing back to the neutral (N) of the service transformer through ground will be picked up by the relay through the current transformer.

Unlike **core balance** or **residual** ground fault protection, **ground return** ground fault protection connects directly to the system grounded neutral.

False! **(E) Core balance ground fault protection does NOT connect directly to the system grounded neutral, ground return fault protection does.**

The answer is: **A and B.**

(Multiple correct AIT question) - Ch. 9.2 Protective Relaying

Problem #62 Solution

ANSI #86 is a lockout relay master trip device. This is a high-speed auxiliary contact relay that is used as an intermediary device between protection relays and circuit breakers to trip or isolate more than one circuit. By itself, it is NOT a protection relay.

> False! **(A) ANSI #86 is NOT a protection relay.**

> True! **(B) ANSI #86 is a multicontact relay used as an intermediary device between protection relays and circuit breakers to trip or isolate more than one circuit.**

Lockout relays are used with protection relays when more than one breaker needs to be tripped from the same protection relay for a single event, such as transformer differential protection when both the primary and secondary transformer breakers are tripped, bus protection when the main breaker and all feeder breakers are tripped, or generator protection when a large amount of auxiliary equipment, like cooling water pumps and boiler feed water pumps, needs to be quickly de-energized along with the generator.

Lockout relays have a large number of high-speed normally open (N.O.) and normally closed (N.C.) contacts that change state when the lockout relay coil is energized by the trip signal of a protection relay. The normally open contacts are used to energize the trip coils (TC) of circuit breakers and the normally closed contacts are used to interlock control circuits to prevent equipment from energizing or restarting. For example, a normally closed contact can be used in series with the start circuit for a motor so that when the lockout relay is energized, the normally closed contact opens, de-energizing the motor start circuit and preventing the motor from running.

Unlike traditional electrical relays whose contacts revert back to their default state when the relay coil is de-energized (like a motor starting relay or a pilot relay), the ANSI #86 lockout relay must be physically reset even after the relay coil is no longer energized. This is to ensure that the equipment wired to the lockout relay cannot be accidentally restarted.

> False! **(C) ANSI #86 DOES NOT HAVE a built-in automatic reset so that tripped circuit breakers may reclose after a fault.**

The only input to the coil of the ANSI #86 lockout relay is the output trip signal of a protection relay. It does not receive current or voltage inputs from current transformers or potential transformers. Unlike protection relays that compare current, voltage, or frequency inputs to a specified trip threshold, the lockout relay is a simple device composed of one relay coil and multiple normally open (N.O.) and normally closed (N.C.) contacts that change state when the relay coil is energized.

False! **(D) ANSI #86 DOES NOT HAVE one current transformer and one potential transformer as the minimum input.**

False! **(E) The ANSI #86 device does NOT detect sensitive ground faults, or any type of fault.**

The answer is: **B.**

(Multiple correct AIT question) - Ch. 9.3 Protective Relaying

Problem #63 Solution

The most common application for the ANSI #21 distance relay is protection of transmission lines. Distance relays need at a minimum one current input from a current transformer and one voltage input from a potential transformer for each phase of the transmission line being monitored and protected.

By monitoring the voltage and current on the line, the distance relay is able to calculate the impedance seen by the system at any given time using Ohm's law ($V = IZ$). Distance relays help to coordinate the transmission line to ensure that the transmission line circuit breaker is only tripped for faults that occur within a specified distance of the transmission line by comparing the calculated impedance from the current and voltage inputs to the set impedance, which is usually a factor of the transmission line's zero sequence impedance.

In each of the phasor diagrams shown below, the horizontal axis stands for resistance (R) and the vertical axis stands for reactance (X).

Reactance distance relay:

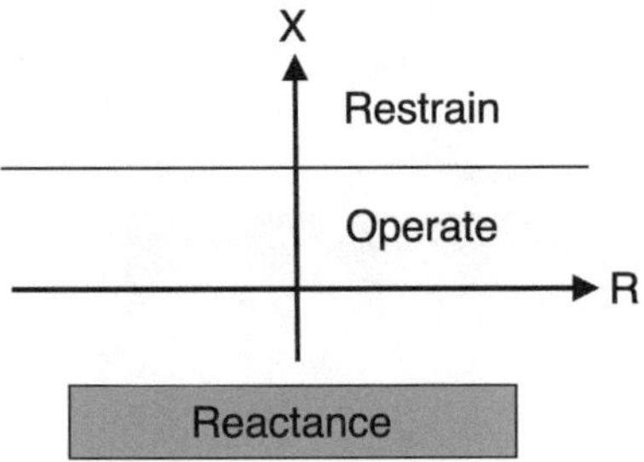

The **reactance** distance relay is one of the most basic types of distance relay. It compares the calculated reactance (X) on the transmission line to a specific reactance setpoint.

When the calculated reactance (X) falls below the setpoint, the reactance distance relay will operate, sending a trip signal to the connected circuit breaker. In the impedance diagram above, the single horizontal line represents the reactance setpoint. This is a non-directional distance relay, meaning it can trip for faults occurring behind the line being protected.

Mho supervised distance relay:

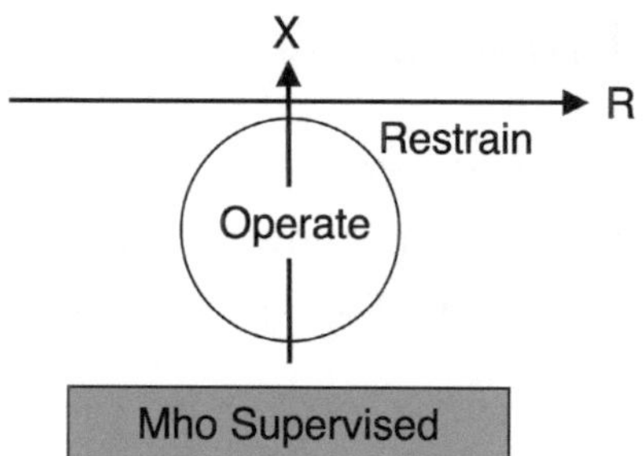

The **mho supervised** distance relay is a combination of a reactance relay and a mho distance relay. The reactance function being supervised by the mho function makes this an overall directional relay that will only trip for faults located in the line being protected.

Impedance distance relay:

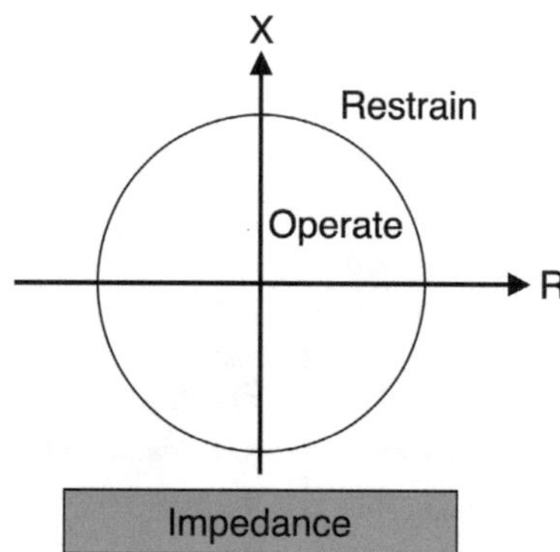

The **impedance** distance relay measures and compares the magnitude of impedance. This is a non-directional distance relay, meaning it can trip for faults occurring behind the line being protected.

Mho distance relay:

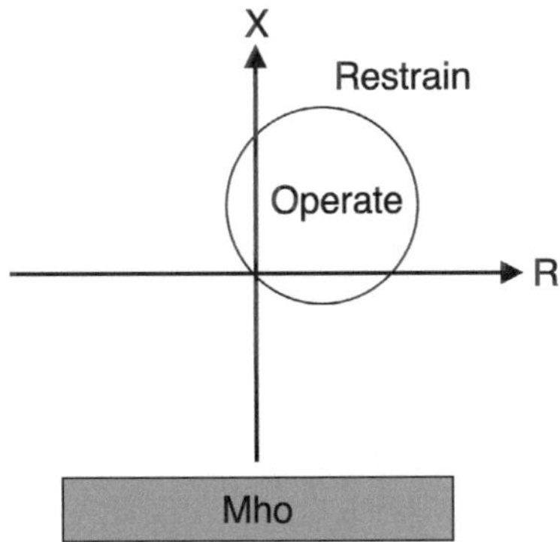

The **mho** distance relay is the directional version of an impedance relay. It measures and compares admittance.

Directional impedance distance relay:

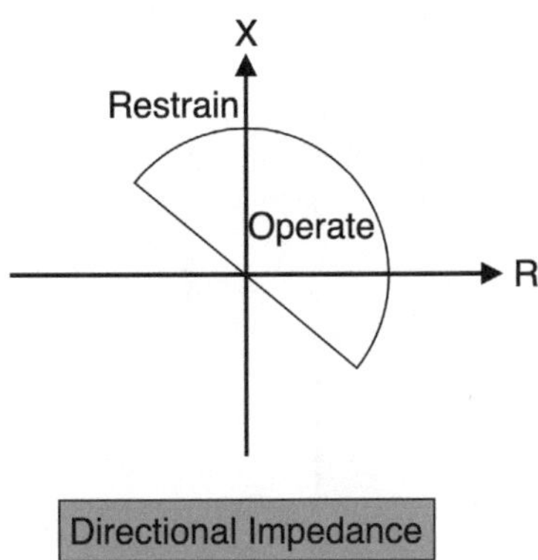

The **directional impedance** distance relay combines an impedance relay and a directional current relay to make the impedance relay directional so that it only trips for faults located on the line being protected.

Here is the completed drag and drop diagram:

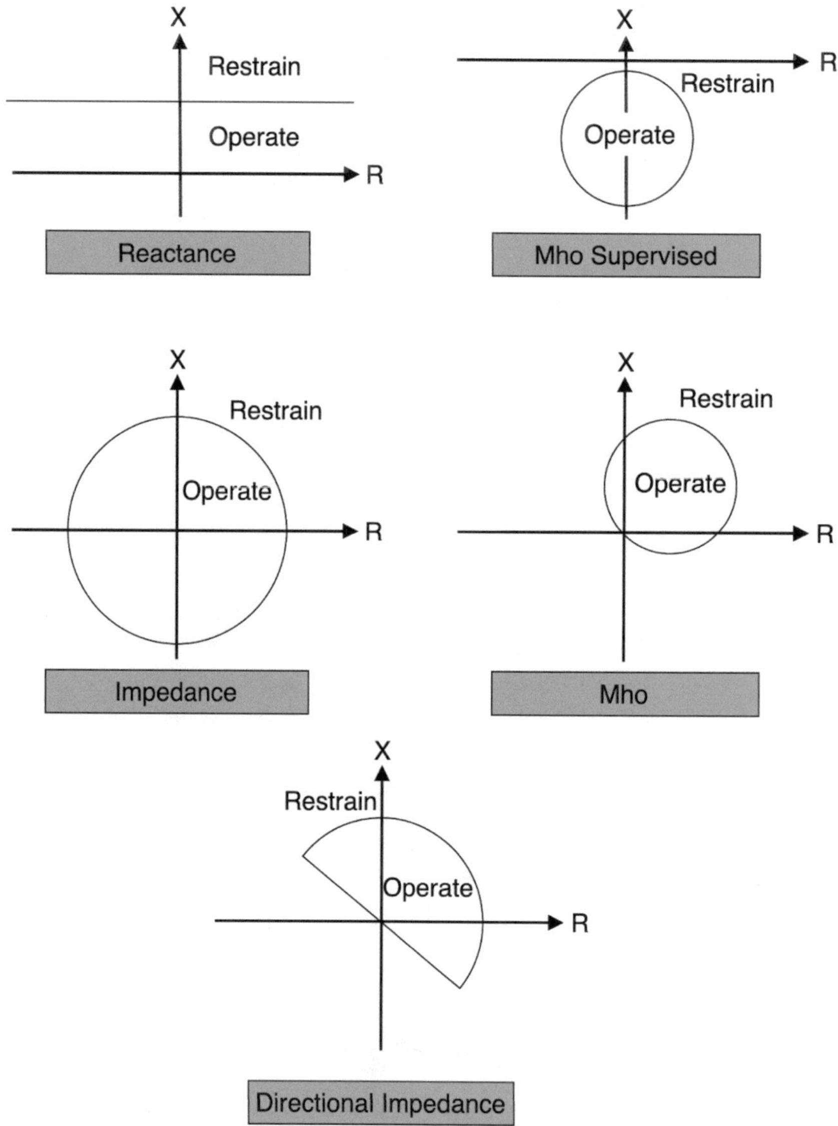

The answer is: **See completed drag and drop diagram above.**

(Drag and drop AIT question) - Ch. 9.2 Protective Relaying

Problem #64 Solution

Selecting the proper fuse for each specific type of application is crucial for proper protection and coordination. Let's evaluate each possible statement to identify which apply to each low-voltage fuse class type.

(A) **Class H fuses are intended for current limiting applications** and are available in 250 volt or 600 volt ratings up to 600 amps.

Class H fuses are rated for 250 volt or 600 volt applications for 600 amps or less with an interrupting rating of 10 kA. However, they are **not** current limiting and should not be used in current limiting applications.

False! **(A) Class H fuses are NOT intended for current limiting applications rated for 250 volts or 600 volts.**

(B) **Class R, Class J, and Class L fuses are intended for current limiting applications.**

Class R fuses are **current limiting** and rated up to 600 amps, 250 volts and 600 volts, with an interrupting rating of 200 kA or 300 kA, and are available in fast acting or time-delay.

Class J fuses are **current limiting** and rated up to 600 amps, with an interrupting rating of 200 kA for up to 600 volts, and are available in fast acting or time-delay.

Class L fuses are **current limiting** and rated from 601 amps to 6,000 amps, 600 volts, with an interrupting rating of 200 kA or 300 kA.

True! **(B) Class R, Class J, and Class L fuses are intended for current limiting applications.**

(C) **Class H and Class K fuses are interchangeable in physical dimensions, but Class H has greater interrupting ratings.**

Class K fuses are manufactured in dimensions interchangeable with class H fuses. While class H fuses have an interrupting rating of 10 kA, class K fuses have an interrupting rating of 50 kA, 100 kA, and 200 kA.

False! **(C) Class H and Class K fuses are interchangeable in physical dimensions, but Class H DOES NOT have greater interrupting ratings.**

(D) Dual-element time-delay fuses may be used to protect against short circuit overcurrents while permitting normal temporary overloads.

Dual-element fuses have an overload fusible link and a short circuit fusible link in series. The overload link allows normal and expected temporary overloads to occur for a specified amount of time without melting the fuse and interrupting the circuit. The short circuit link protects against dangerous fault current. Dual-element time delay fuses are common for motors, transformers, and other devices with high inrush current.

True! **(D) Dual-element time-delay fuses may be used to protect against short circuit overcurrents while permitting normal temporary overloads.**

(E) Current limiting fuses may be used to protect downstream circuit breakers that have an interrupting rating less than the available short circuit current.

Current limiting fuses are extremely fast acting fuses that limit the amount of fault current.

In applications where the interrupting rating of circuit breakers is less than the available short circuit current, current limiting fuses may be used to decrease the amount of available short circuit current to levels that are within range of the circuit breaker rating.

True! **(E) Current limiting fuses may be used to protect downstream circuit breakers that have an interrupting rating less than the available short circuit current.**

The answer is: **B, D, and E.**

(Multiple correct AIT question) - Ch. 9.3 Protective Devices

Problem #65 Solution

Molded-case circuit breakers (MCCBs) and low-voltage power circuit breakers (LVPCBs) are the two types of low-voltage circuit breakers. Insulated-case circuit breakers (ICCBs) are a specific subtype of the molded-case circuit breaker (MCCB).

Let's evaluate each possible statement to identify which apply to each type.

(A) Molded-case circuit breakers are fast interrupting circuit breakers with all components built into a single insulating molded case that are mostly fixed-mounted and typically operated with a mechanical toggle.

Molded-case circuit breakers (MCCBs) are totally enclosed with all of their components built into a single insulated molded case. They are most often fixed-mounted directly to the panel. These are the least expensive type of circuit breakers and are rated for 80% of their continuous current rating unless specified otherwise since they are not operated in free air.

Most molded-case circuit breakers (MCCBs) are manually operated with a mechanical toggle switch built directly into the frame and with the continuous current rating written directly on the toggle switch.

True! **(A) Molded-case circuit breakers are fast interrupting circuit breakers with all components built into a single insulating molded case that are mostly fixed-mounted.**

(B) Insulated-case circuit breakers are larger in frame size compared to molded-case circuit breakers and feature a stored energy operating mechanism for operation.

The insulated-case circuit breaker (ICCB) is a more expensive, typically heavier duty subtype of the molded-case circuit breaker (MCCB).

The biggest difference between the two is that while the molded-case circuit breaker (MCCB) is typically manually operated with a mechanical toggle switch built directly into the frame, the insulated-case circuit breaker (ICCB) uses a built-in stored energy mechanism to operate the open/close/trip operation of the breaker such as a spring that must be manually or electrically "charged" before use.

True! **(B) Insulated-case circuit breakers are larger in frame size compared to molded-case circuit breakers and feature a stored energy operating mechanism for operation.**

(C) Insulated-case circuit breakers are most commonly used for residential and commercial lighting and receptacle circuits.

The miniature circuit breaker (MCB), which is also a subtype of the molded-case circuit breaker (MCCB), is most commonly used for residential and commercial lighting and receptacle circuits, not the insulated-case circuit breaker (ICCB).

The miniature circuit breaker (MCB) is the common type of 120 volt, 240 volt, or 208V volt circuit breaker used in 120 volt or 208 volt panels.

False! **(C) Insulated-case circuit breakers are NOT most commonly used for residential and commercial lighting and receptacle circuits.**

(D) Low-voltage power circuit breakers are larger in frame size compared to Insulated-case circuit breakers and primarily used in drawout switchgear.

The low-voltage power circuit breaker (LVPCB) is the largest low-voltage circuit breaker frame size. Unlike molded-case circuit breakers, low-voltage power circuit breakers (LVPCBs) are built with open construction for maintenance purposes.

Instead of being fixed-mounted, low-voltage power circuit breakers (LVPCBs) are connected and disconnected to the bus by a drawout mechanism. This allows for easier removal of the circuit breaker for maintenance purposes and also improves safety by being able to visibly verify that the circuit is open.

Unlike switchboards, which may use a variety of overcurrent protection devices, switchgear may only use low-voltage power circuit breakers (LVPCBs), which are the most expensive type of low-voltage circuit breaker.

True! **(D) Low-voltage power circuit breakers are larger in frame size compared to Insulated-case circuit breakers and primarily used in drawout switchgear.**

(E) Low-voltage power circuit breakers are rated to interrupt 100% of their continuous current rating.

Unlike insulated-case circuit breakers (ICCBs) and molded-case circuit breakers (MCCBs) that are rated for 80% of their continuous current rating unless otherwise specified, the low-voltage power circuit breaker (LVPCB) are rated up to 100% of their continuous current rating within their manufactured enclosure.

True! **(E) Low-voltage power circuit breakers are rated to interrupt 100% of their continuous current rating.**

The answer is: **A, B, D, and E.**

(Multiple correct AIT question) - Ch. 9.3 Protective Devices

Problem #66 Solution

The majority of current transformer problems (if not all) on the PE exam will be ideal current transformers where **percent error** is not a factor and the current transformer excitation current (I_e) is ignored. In the real world, current transformers are wound around an iron core that draws an excitation current (I_e) and is subject to **magnetic saturation** if too much current is drawn.

When dealing with non-ideal current transformers, the amount of error introduced to the relay current signal gets worse as the excitation current (I_e) in the current transformer increases.

According to the problem, **the ANSI #50P relay is wired to the X3 and X4 MRCT taps**, which means the MRCT is wired for a partial CT ratio of **250:5**:

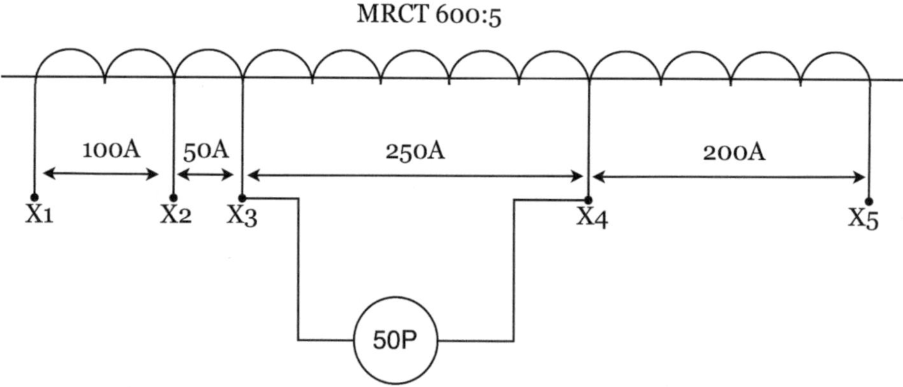

Also according to the problem, the **relay tap setting is 10 amps**, which means the relay is programmed to trip when 10 amps flow through it (I_r = 10A):

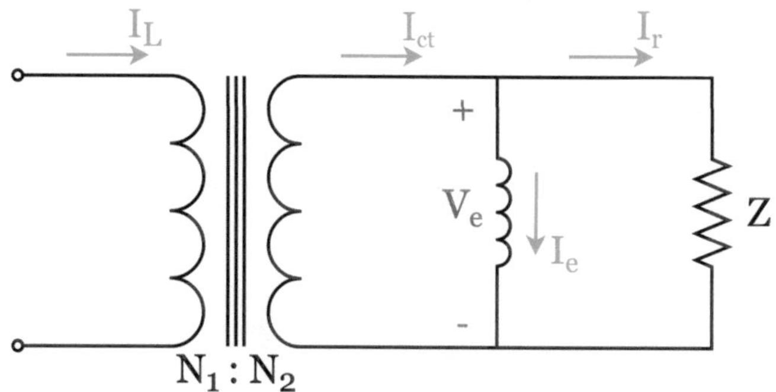

Current Transformer Equivalent Circuit

where:

I_L = line current [A]
N_1:N_2 = CT turns ratio

I_{CT} = secondary CT current [A]
V_e = excitation voltage [V]
I_e = excitation current [A]
I_r = relay current [A]
Z = total burden [Ω] in the CT circuit

To determine the **minimum fault current** (I_L) on the line that will result in 10 amps through the relay (I_r), we first need to use the graph to determine the amount of excitation current (I_e) flowing through the CT core when the core is excited with **50 volts** (V_e):

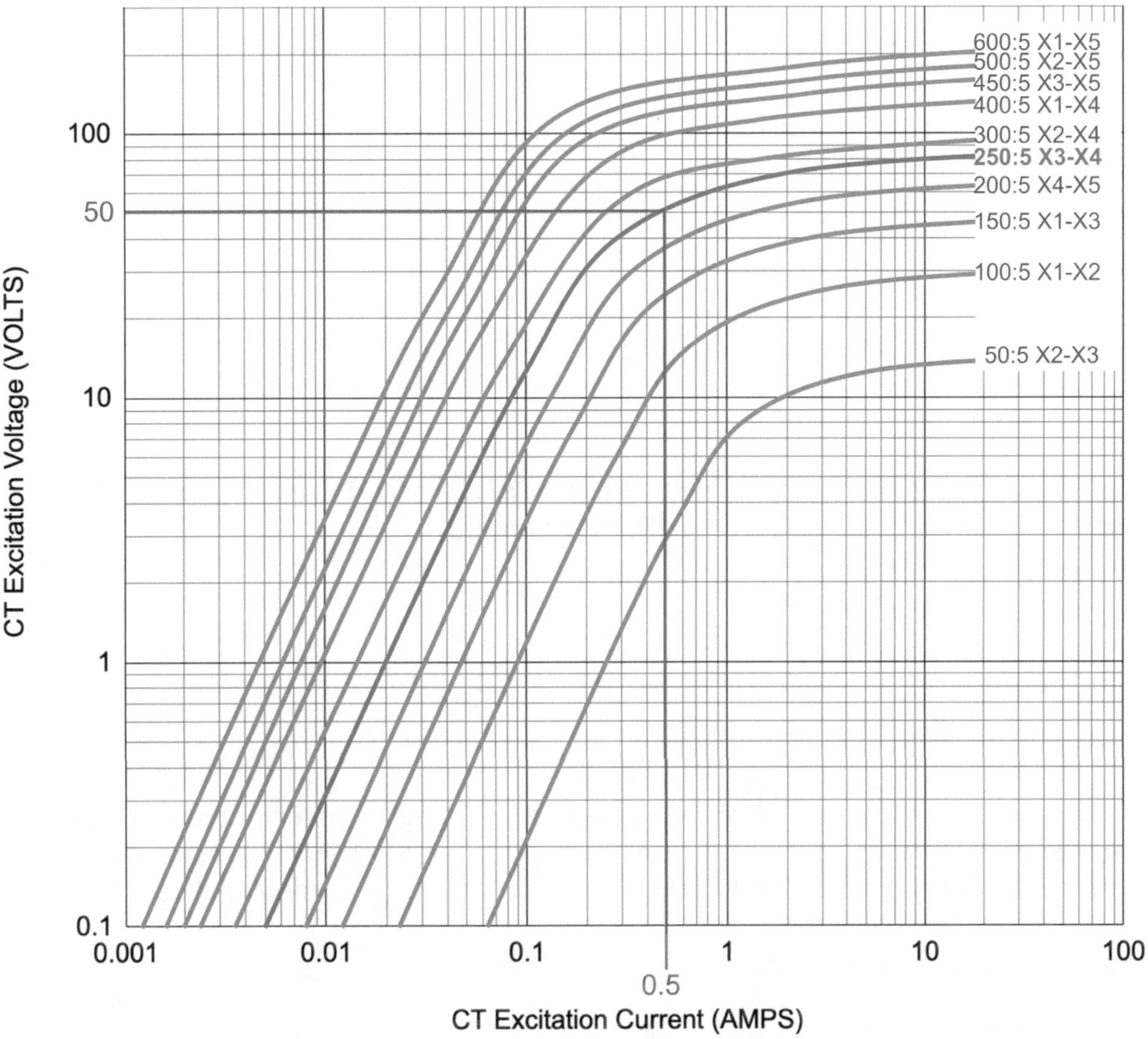

600:5 X1-X5
500:5 X2-X5
450:5 X3-X5
400:5 X1-X4
300:5 X2-X4
250:5 X3-X4
200:5 X4-X5
150:5 X1-X3
100:5 X1-X2
50:5 X2-X3

CT Excitation Voltage (VOLTS): 0.1, 1, 10, 50, 100
CT Excitation Current (AMPS): 0.001, 0.01, 0.1, 0.5, 1, 10, 100

Using the graph from the problem above and counting in logarithmic scale, **50 volts of excitation voltage** (V_e) across the core results in **0.5 amps of excitation current** (I_e) when the MRCT is wired for a **250:5 ratio** across the **X3 and X4** MRCT taps.

Using the current transformer equivalent circuit, we can calculate the secondary CT current (I_{ct}) using Kirchoff's current law:

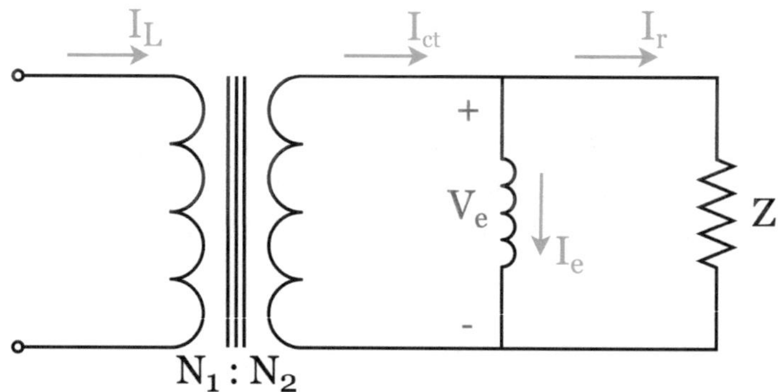

Current Transformer Equivalent Circuit

$$I_{ct} = I_e + I_r$$

$$I_{ct} = 0.5\,A + 10\,A$$

$$I_{ct} = 10.5\,A$$

Next, we can use the CT current ratio to step up the secondary CT current (I_{ct}) through the current transformer to calculate the resulting line current (I_L):

$$I_L = 10.5\,A\left(\frac{250}{5}\right)$$

$$I_L = 525\,A$$

This means that for 10 amps to flow through the relay (I_r) resulting in relay operation, **there must be a minimum of 525 amps on the line** (I_L). Any value less than 525 amps of line current will result in less than the 10 amp relay tap setting and the relay will not operate.

Out of the possible answer choices:

(A) 600 amps, (B) 550 amps, (C) 500 amps, (D) 450 amps, and (E) 400 amps

Only answer choices **(A) 600 amps and (B) 550 amps** of fault current on the line (I_L) will result in relay operation. The remaining choices result in less than 10 amps of relay current.

*Note: had we assumed an **ideal** current transformer and neglected the excitation voltage vs excitation current graph in the problem, we would have erroneously included answer choice (C) 500 amps since 500 amps of line current (I_l) will result in 10 amps of relay current when the excitation current is neglected (I_e = 0).*

*Note: if you are looking for practice calculating CT percent error, please see problem #31 from the **Electrical Engineering PE Practice Exam and Technical Study Guide**, also by Zach Stone, P.E.*

The answer is: **A and B.**

(Multiple correct AIT question) - Ch. 9.3 Protective Devices

Problem #67 Solution

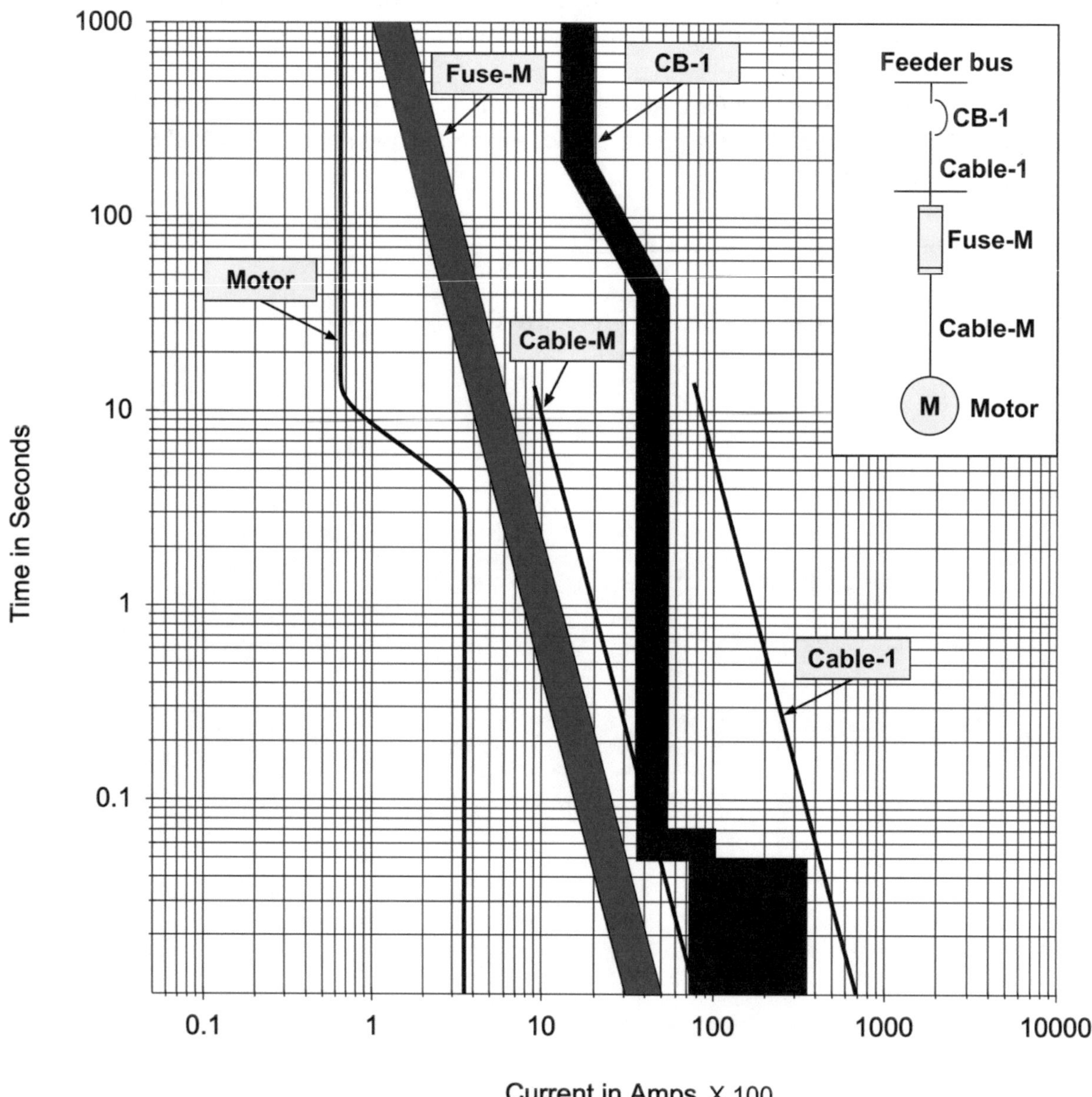

From left to right shown on the TCC graph above:

Motor:

This is the **motor-starting and capability curve** for the motor on the single-line diagram. This curve represents the initial **locked rotor starting current** and the **full load steady state current** of the motor once it has reached rated operating speed.

Fuse-M:

This is the **device trip characteristics** for the fuse labeled "Fuse-M" on the single-line diagram. Notice that the fuse curve is completely to the right and above the motor starting and capability curve with no overlap. This ensures that the motor will not trip the fuse during motor startup and while operating at rated conditions.

Cable-M:

This is the **conductor damage curve** for the conductor labeled "Cable-M" on the single-line diagram. Notice that this curve is completely to the right and above the trip characteristics for Fuse-M. This ensures that the fuse will trip and clear any fault current before it has the chance to damage Cable-M.

CB-1:

This is the **device trip characteristics** for the circuit breaker labeled "CB-1" on the single-line diagram. Notice there is no overlap with the trip characteristics for Fuse-M, which ensures that Fuse-M will trip first, clearing any downstream faults before CB-1 will operate.

Cable-1:

This is the **conductor damage curve** for the conductor labeled "Cable-1" on the single-line diagram. Notice that this curve is completely to the right and above the trip characteristics for CB-1. This ensures that the circuit breaker will trip and clear any fault current before it has the chance to damage Cable-1.

Cable-1 is a **feeder** conductor that is upstream of the **branch** conductor Cable-M. This means that Cable-1 is likely rated at a higher ampacity compared to Cable-M and would also have a damage curve rated at a higher fault current level, which is why it is to the right of Cable-M's damage curve on the graph.

The answer is: **See completed drag and drop diagram shown above.**

(Multiple correct AIT question) - Ch. 9.4 Coordination

Problem #68 Solution

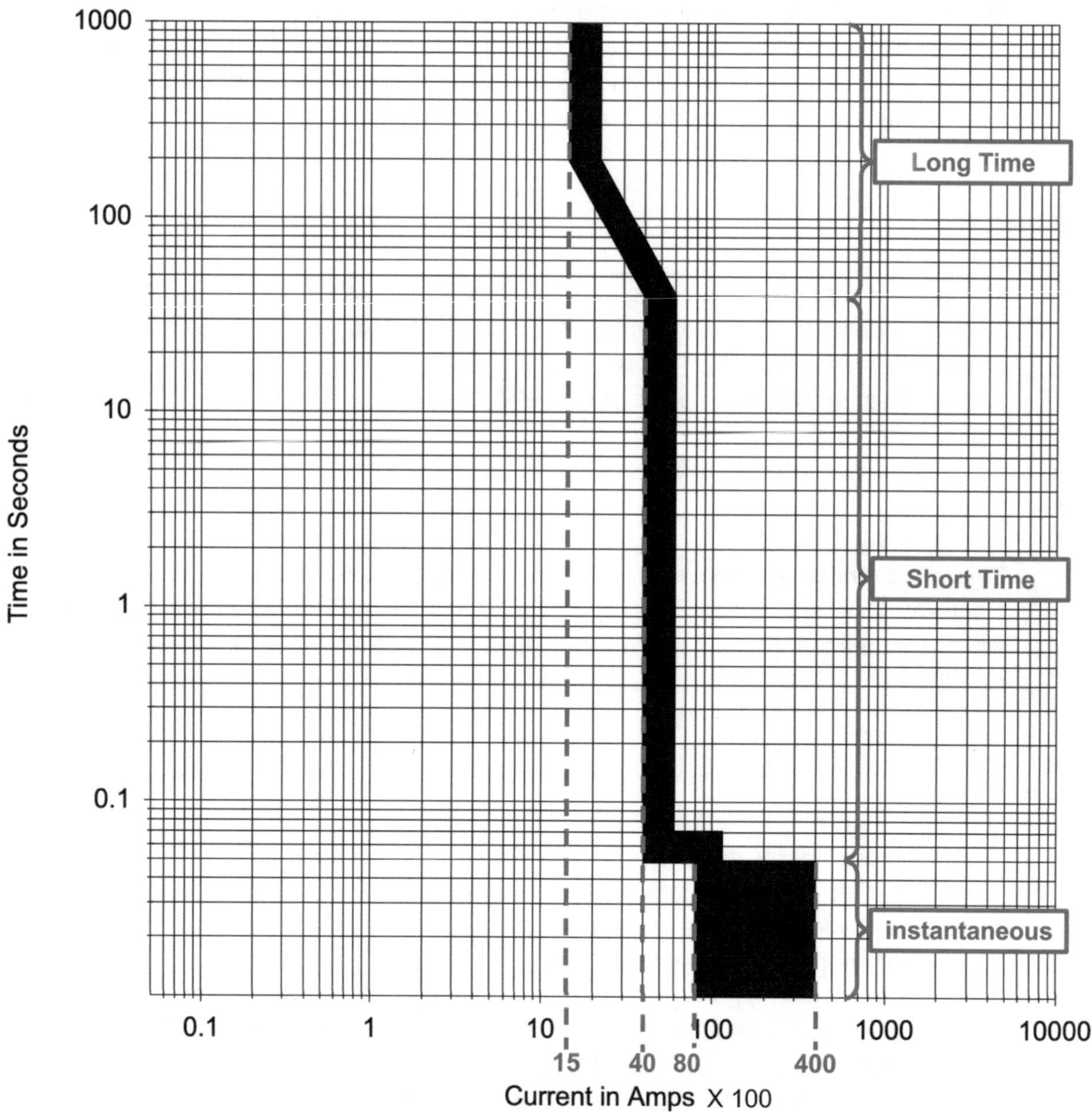

We will need to be familiar with the different types of time overcurrent trip characteristics that appear on a time current characteristic (TCC) curve as well as count in logarithmic scale to answer this problem correctly.

> ***Careful!*** *The horizontal axis is not only in logarithmic scale, but it is also scaled in multiples of 100 ("X 100"). This means that we will* ***need to multiply the pickup current values shown on the graph by 100*** *to calculate their true values.*

From left to right on the time current characteristic (TCC) curve:

The **long-time pickup** is **1,500 amps.**

The **short-time pickup** is **4,000 amps.**

The **instantaneous pickup** is **8,000 amps.**

The **maximum interrupting current** is **40,000 amps**.

The long-time and short-time trip regions are **ANSI #51 inverse time overcurrent** relay functions, and the instantaneous trip region is the **ANSI #50 instantaneous overcurrent** relay function[94].

The answer is: **1,500 amps, 4,000 amps, 8,000 amps, and 40,000 amps.**

(Multiple correct AIT question) - Ch. 9.4 Coordination

[94] *NCEES® Reference Handbook (Version 1.1.2) - 3.1.6 Single-Line Diagrams p. 38*

Problem #69 Solution

The graph in the problem shows two of the three main types of inverse time curves that can be applied to most short-time and long-time trip characteristics of ANSI #51 inverse time overcurrent relays[95]:

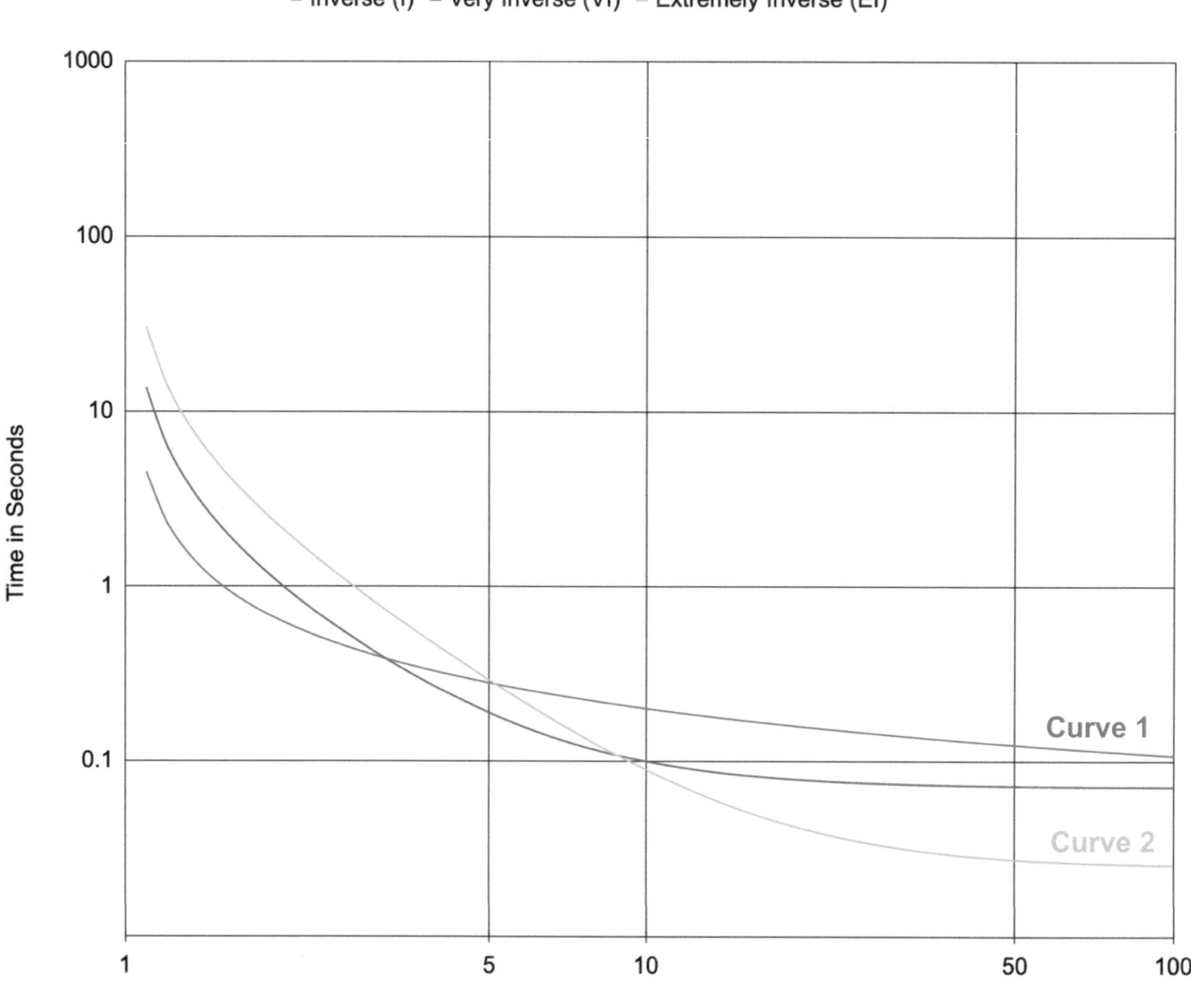

Curve 1 is the **inverse (I)** time curve, curve 2 is the **extremely inverse (EI)** time curve, and not shown on the graph in the problem is the **very inverse (VI)** time curve.

Let's evaluate each of the possible statements from the problem to see which ones apply to curve 1 and curve 2:

---

95 *NCEES® Reference Handbook (Version 1.1.2) - 3.1.6 Single-Line Diagrams p. 38*

(A) Curve 1 is more time inverse compared to curve 2.

The more inverse a time current characteristic (TCC) curve is, the **faster** it will trip for larger magnitudes of fault current. Even though the curves are not labeled in the problem, we can tell that curve 1 is **less** time inverse compared to curve 2 since at larger fault current it will take more time to operate.

For example, see how curve 1 takes longer to operate compared to curve 2 at 50 multiples of pickup:

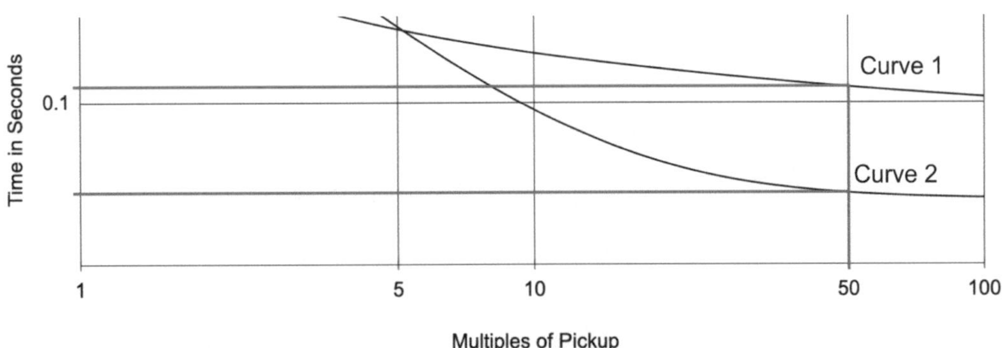

False! **(A) Curve 1 is NOT more time inverse compared to curve 2.**

(B) Curve 2 is more time inverse compared to curve 1.

As we learned while evaluating answer choice (A), curve 2 is more time inverse than curve 1 since curve 2 will trip faster at higher levels of fault current compared to curve 1.

True! **(B) Curve 2 is more time inverse compared to curve 1.**

(C) Both curves are examples of ANSI #51 relay functions.

When programing ANSI #51 inverse time overcurrent relays, there are generally three types of adjustable settings for both the long- and short-time trip regions. They are **pickup current** (also known as the tap setting), **time delay** (also known as time dial), and the type of **inverse time curve**.

Changes in **pickup** shift the curve along the **horizontal current axis**.

Changes in **time delay** shift the curve along the **vertical time axis**. For example, compare the difference in the very inverse (VI) time curve from a time delay setting of 0.5 all the way to 10:

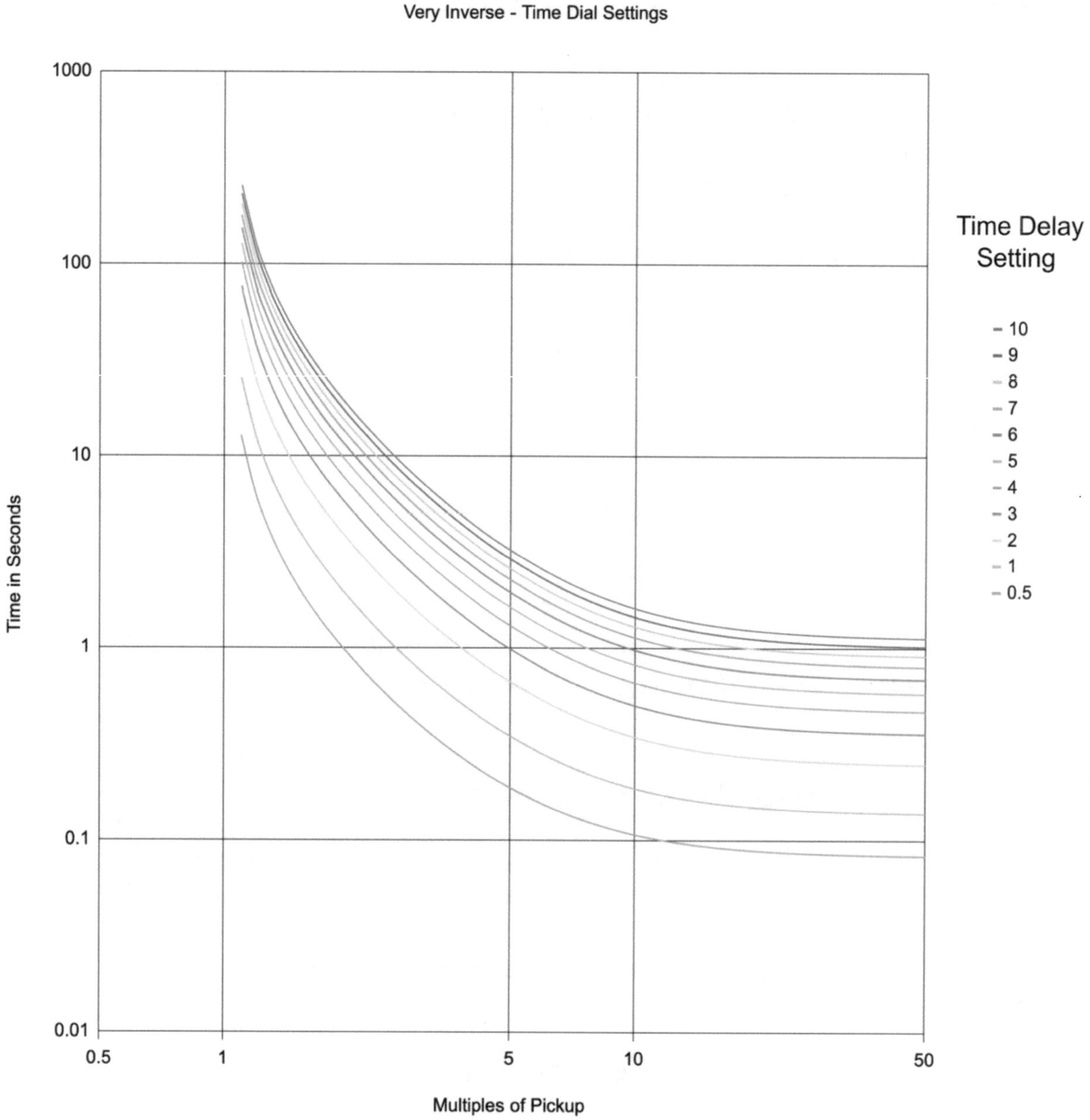

Differences in Time Delay

Changes in the **inverse setting** of the time curve change the **slope** of the curve. For example, compare the difference in slope between the inverse (I) time curve, the very inverse (VI) time curve, and the extremely inverse (EI) time curve:

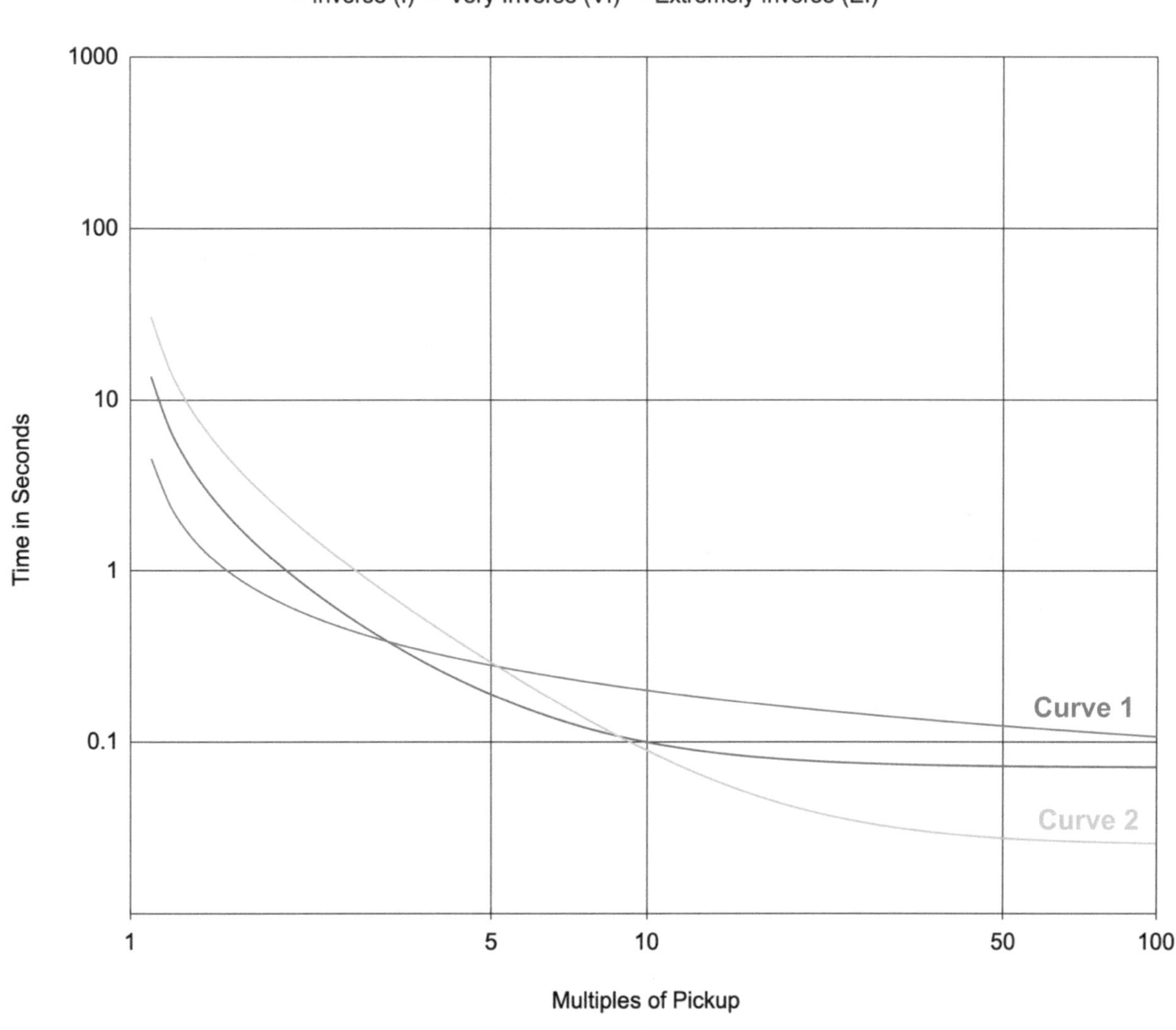

Differences in Inverse Curve Type

True! **(C) Both curves are examples of ANSI #51 relay functions.**

(D) Both curves are examples of ANSI #50 relay functions.

The ANSI #50 relay function is **instantaneous time overcurrent** which has a flat horizontal non-adjustable slope.

The curves in the problem are for ANSI #51 inverse time overcurrent, not ANSI #50 instantaneous time overcurrent.

False! **(D) Both curves are NOT examples of ANSI #50 relay functions.**

(E) Curve 1 and curve 2 are examples of different time dial settings.

Increasing or decreasing the time dial setting of an overcurrent relay results in an increase or decrease in the amount of time before the relay will trip for the same level of fault current.

This results in shifting the curve up (more time, increase) or down (less time, decrease) along the vertical time axis. It does **not** change the slope of the curve.

False! **(E) Curve 1 and curve 2 are NOT examples of different time dial settings.**

The answer is: **B and C.**

(Multiple correct AIT question) - Ch. 9.4 Coordination

Problem #70 Solution

Hazardous classified locations in the *2017 National Electrical Code®* are defined in ***NEC®* Article 500**. The presence of different types of hazards must follow special electrical design considerations according to the *NEC®*, so it is important to be able to recognize each hazard class type. In general, the **three different classes** are locations with:

Class I - Flammable gases, flammable vapors, and combustible vapors.

Class II - Combustible dust.

Class III - Ignitible fibers.

Each of the three classes are separated into two different **severity divisions**. Each division has its own unique definition depending on its class, but in general they are:

Division 1 - Potential for combustion or explosion during normal operations, normal maintenance, or normal manufacturing, or due to equipment failure.

Division 2 - Storage or handling with the potential for combustion or explosion only during storage failure.

Let's evaluate each of the hazardous classifications in the answer choices to determine the appropriate class and division type:

(A) A textile mill that produces fabric containing cotton fibers that are easily ignitible.

NEC® 500.5(D) Class III Locations - hazardous because of the presence of easily ignitable fibers.

NEC® 500.5(D)(1) Class III, Division 1. - location in which easily ignitable fibers/flyings are handled, manufactured, or used.

This is an example of a **Class III Division 1** hazardous location.

Class III Division 1

(B) A grain milling facility that produces combustible dust with the potential of explosion only if the grain milling machinery fails.

NEC® 500.5(C) Class II Locations - hazardous because of the presence of combustible dust.

NEC® 500.5(C)(1) Class II, Division 1. (2) - where mechanical failure or abnormal operation of machinery or equipment might cause such explosive or ignitible mixtures to be produced, and might also provide a source of ignition through simultaneous failure of electrical equipment.

This is an example of a **Class II Division 1** hazardous location.

Class II Division 1

(C) A textile storage facility that stores and transports cotton fibers that are easily ignitible.

NEC® 500.5(D) Class III Locations - hazardous because of the presence of easily ignitable fibers.

NEC® 500.5(D)(2) Class III, Division 2. - location in which easily ignitable fibers/flyings are stored or handled other than in the process of manufacture.

This is an example of a **Class III Division 2** hazardous location.

Class III Division 2

(D) A dry cleaning plant where combustible liquid-produced vapors are present in quantities sufficient to produce a combustion only when the cleaning machinery is undergoing routine maintenance.

NEC® 500.5(B) Class I Locations - flammable gases, flammable liquid–produced vapors, or combustible liquid–produced vapors are or may be present in the air in quantities sufficient to produce explosive or ignitible mixtures.

NEC® 500.5(B)(1) Class I, Division 1. (2) - combustible liquids above their flash points may exist frequently because of repair or maintenance operations or because of leakage.

This is an example of a **Class I Division 1** hazardous location.

Class I Division 1

(E) A grain processing facility that handles combustible dust that settles over time on top of electrical equipment which, if not routinely cleaned, causes 480 volt power and lighting transformers to overheat.

NEC® 500.5(C) Class II Locations - hazardous because of the presence of combustible dust.

NEC® 500.5(C)(2) Class II, Division 2. (3) - in which combustible dust accumulations on, in, or in the vicinity of the electrical equipment could be sufficient to interfere with the safe dissipation of heat from electrical equipment, or could be ignitible by abnormal operation or failure of electrical equipment.

This is an example of a **Class II Division 2** hazardous location.

Class II Division 2

(F) A chemical storage facility that receives, stores, and transports dry cleaning liquids in enclosed airtight containers. The liquid chemicals produce combustible vapors in sufficient quantities that would result in combustion if the containers were to fail.

NEC® 500.5(B) Class I Locations - flammable gases, flammable liquid–produced vapors, or combustible liquid–produced vapors are or may be present in the air in quantities sufficient to produce explosive or ignitible mixtures.

NEC® 500.5(B)(2) Class I, Division 2. (1) - combustible liquid–produced vapors are handled, processed, or used, but in which the liquids, vapors, or gases will normally be confined within closed containers or closed systems from which they can escape only in case of accidental rupture or breakdown of such containers.

This is an example of a **Class I Division 2** hazardous location.

Class I Division 2

Here is the completed drag and drop diagram:

(A) A textile mill that produces fabric containing cotton fibers that are easily ignitible.

Class III Division 1

(B) A grain milling facility that produces combustible dust with the potential of explosion only if the grain milling machinery fails.

Class II Division 1

(C) A textile storage facility that stores and transports cotton fibers that are easily ignitible.

Class III Division 2

(D) A dry cleaning plant where combustible liquid-produced vapors are present in quantities sufficient to produce a combustion only when the cleaning machinery is undergoing routine maintenance.

Class I Division 1

(E) A grain processing facility that handles combustible dust that settles over time on top of electrical equipment which if not routinely cleaned causes 480 volt power and lighting transformers to overheat.

Class II Division 2

(F) A chemical storage facility that receives, stores, and transports dry cleaning liquids in enclosed airtight containers. The liquid chemicals produce combustible vapors in sufficient quantities that would result in combustion if the containers were to fail.

Class I Division 2

The answer is: **See completed drag and drop diagram above.**

(Multiple correct AIT question) - Ch. 3.1 National Electrical Code (NEC®)

Problem #71 Solution

Let's evaluate each of the five possible choices to see which ones apply to Class I Division 1 hazardous locations according to the *2017 National Electrical Code®*:

(A) Wall-mounted PVC conduit is permitted for Class I Division 1 hazardous locations.

NEC® 501.10(A)(1)(a) Exception: Type PVC shall be permitted where encased in a concrete envelope a minimum of 50 mm (2 in.) thick and provided with not less than 600 mm (24 in.) of cover measured from the top of the conduit to grade.

PVC is permitted for Class I Division 1 hazardous locations, but it must be encased in concrete according to the above specifications. It may not be wall mounted.

False! **(A) Wall-mounted PVC conduit is NOT permitted for Class I Division 1 hazardous locations.**

(B) Wiring methods that are allowable for Class I Division 1 hazardous locations are also permitted for use in Class I Division 2 hazardous locations.

NEC® 501.10(B)(1) General - In Class I Division 2 locations, all wiring methods permitted in 501.10(A) shall be permitted.

NEC® 501.10(A) are the rules for Class I Division 1 hazardous locations.

In other words, if a wiring method is acceptable for Class I Division 1, which is more conservative than Class I Division 2, then that method is also acceptable for Class I Division 2.

True! **(B) Wiring methods that are allowable for Class I Division 1 hazardous locations are also permitted for use in Class I Division 2 hazardous locations.**

(C) Hermetically sealed electrical enclosures do not require conduit seals in Class I Division 1 hazardous locations.

Conduit seals are generally required in Class I Division 1 for two reasons. The first reason is to prevent an explosion inside one conduit or enclosure from spreading to another conduit or enclosure. The second reason for conduit seals is to prevent combustible or flammable gasses and vapor that are present in Class I Division 1 hazardous locations from spreading to nonhazardous locations.

However, there is an exception to this if the electrical enclosure is hermetically sealed:

NEC® 501.15(A)(1)(1) Exception: Seals shall not be required for conduit entering an enclosure under any one of the following conditions: a. The switch, circuit breaker, fuse, relay, or resistor is enclosed within a chamber hermetically sealed against the entrance of gases or vapors.

The *National Electrical Code*® defines **hermetically sealed** in NEC® Article 100 - Definitions as: Equipment sealed against the entrance of an external atmosphere where the seal is made by fusion, for example, soldering, brazing, welding, or the fusion of glass to metal.

True! **(C) Hermetically sealed electrical enclosures do not require conduit seals in Class I Division 1 hazardous locations.**

(D) General-purpose electrical enclosures may be used for surge-protective devices in Class I Division 1 hazardous locations as long as they are nonarcing.

Even if the surge-protective device is nonarcing, it must still be installed in a Class I Division 1 identified enclosure:

NEC® 501.35 Surge Protection. (A) Class I, Division 1. surge-protective devices shall be installed in enclosures identified for Class I, Division 1 locations.

For the enclosure to be identified for Class I Division 1, it must meet all the requirements of NEC® 500.8(B).

Although general-purpose enclosures are permitted for nonarcing surge-protective devices in Class I Division 2 hazardous locations according to NEC® 501.35(B), this does not meet the requirements for Class I Division 1.

False! **(D) General-purpose electrical enclosures may NOT be used for surge-protective devices in Class I Division 1 hazardous locations EVEN IF they are nonarcing.**

(E) Only 90°C temperature rated conductors are permitted for Class I Division 1 hazardous locations.

The majority of rules for hazardous locations are to prevent combustion and ignition.

A 90°C temperature rated conductor is rated for more ampacity at the same wire size compared to a 60°C and 75°C temperature rated conductor. However, 90°C temperature rated conductors are typically used for ampacity adjustment or correction, or for specific installations where all terminations and devices are also rated for 90°C.

90°C temperature rated conductors would not help prevent combustion or ignition in locations with hazardous levels of gas, dust, or fibers present. There are no rules in NEC® Article 500 or 501 that mandate the use of 90°C temperature rated conductors for Class I Division 1 hazardous locations.

False! (E) 90°C IS NOT THE ONLY temperature rating permitted for conductors Class I Division 1 hazardous locations.

The answer is: **B and C.**

(Multiple correct AIT question) - Ch. 3.1 National Electrical Code (NEC®)

Problem #72 Solution

This is an **ambient temperature conductor ampacity adjustment** problem. Most conductors are rated for either 30°C (86°F) or 40°C (104°F) ambient temperature. When the ambient temperature **exceeds** the ambient temperature rating of the conductor, the amount of current the conductor is rated to safely carry **decreases**.

> *Note: Conductor ambient temperature rating (example: 30°C or 40°C) is not the same as conductor insulation temperature rating (example: 60°C, 75°C, or 90°C).*

THWN is **t**hermoplastic **h**eat- and **w**ater-resistant **n**ylon-coated wire for general-purpose wiring.

There are a handful of different conductor ampacity tables in NEC® Article 310. Since the problem states that we are using **three** copper THWN conductors in **PVC conduit** for a **480 volt** load, the correct conductor ampacity table is NEC® Table 310.15(B)(16):

> NEC® Table 310.15(B)(16) (formerly Table 310.16) Allowable Ampacities of Insulated Conductors Rated Up to and Including 2000 Volts, 60°C Through 90°C (140°F Through 194°F), Not More Than Three Current-Carrying Conductors in Raceway, Cable, or Earth (Directly Buried), Based on Ambient Temperature of 30°C (86°F).
>
> where:
>
> **Formerly Table 310.16** - This is the table number from older versions of the codebook.
>
> **Allowable Ampacity** - The numbers in the table are the ampacity rating of each conductor.
>
> **Insulated Conductors** - These are non-bare conductors that have insulation.
>
> **Rated Up to and Including 2000 Volts** - The ampacity ratings given in the table are good for conductors rated up to a maximum of 2,000 volts.
>
> **60°C Through 90°C (140°F Through 194°F)** - The ampacity ratings given in the table are only for 60°C, 75°C, and 90°C insulation temperature rated conductors.
>
> **Not More Than Three Current-Carrying Conductors** - The ampacity ratings given in the table are the base ampacity ratings without any adjustment factors for up to (no more than) three current-carrying conductors total.
>
> **Raceway, Cable, or Earth (Directly Buried)** - The ampacity ratings given in the table are for conductors that are in raceway, cable, or directly buried in the ground.

> *Note: Conduit is a type of raceway. See NEC® Article 100 - Definitions: Raceway - An enclosed channel designed expressly for holding wires, cables, or busbars, with additional functions as permitted in this Code.*

Based on Ambient Temperature of 30°C (86°F) - The ampacity ratings given in the table are based on ambient temperature conditions of 30° C (86° F) of the surrounding environment where the conductors are to be installed.

Notice that:

According to the title of NEC® Table 310.15(B)(16), the ampacities given in the table are based on an ambient temperature of **30° C** (86° F).

According to the NEC® Table 310.15(B)(16), copper THWN has an insulation temperature rating of **75ºC** (167ºF).

Since the 30ºC ambient temperature rating of NEC® Table 310.15(B)(16) is **less than** the actual 110ºF ambient temperature according to the problem, we will have to apply an **ambient temperature correction factor** according to the *National Electrical Code®.*

There are two ambient temperature correction factor tables in the NEC®:

NEC® Table 310.15(B)(2)(a) Ambient Temperature Correction Factors Based on **30°C** (86°F).

NEC® Table 310.15(B)(2)(b) Ambient Temperature Correction Factors Based on **40ºC** (104ºF)

Since we are using the NEC® Table 310.15(B)(16) conductor ampacity table which is based on an ambient temperature of 30° C (86° F), we will be using the first ambient temperature correction factor table for 30ºC ambient rated conductors: NEC® Table 310.15(B)(2)(a).

According to NEC® Table 310.15(B)(2)(a), the ambient temperature correction factor for 30ºC ambient temperature rated conductors for an actual ambient temperature of 110ºF is **0.82** for 75ºC insulation temperature rated conductors (THWN).

This means that every ampacity rating given in NEC® Table 310.15(B)(16) for THWN conductors (and all other 75ºC insulation temperature rated conductors) must be multiplied (derated) by 0.82 in order to calculate the rated ampacity for use in 110ºF ambient temperature environments.

Let's evaluate each of the possible answer choices using the following table:

THWN Copper			
Wire Size	**30°C (86° F) Ampacity from NEC® Table 310.15(B)(16)**	**Correction Factor for 110°F from NEC® Table 310.15(B)(2)(a)**	**Corrected Ampacity for 110°F**
2 AWG	115 A	0.82	94 A
1 AWG	130 A	0.82	107 A
1/0 AWG	150 A	0.82	123 A
2/0 AWG	175 A	0.82	144 A
3/0 AWG	200 A	0.82	164 A

Notice that the **minimum allowable** THWN copper wire size for 110°F ambient temperature for a 125 amp load in the conditions described in the problem is **2/0 AWG** (2/0 AWG THWN copper is rated for a maximum of 144 amps at 110°F).

Out of the possible THWN copper wire sizes given in the problem, only 2/0 AWG and 3/0 AWG are allowable according to code.

To compare, if the ambient temperature given in the problem was 30°C (86°F) instead of 110°F, then we would **not** need to apply an ambient temperature correction factor to the ampacities from NEC® Table 310.15(B)(16), and the minimum wire size required for THWN copper would instead be 1 AWG (1 AWG THWN copper is rated for a maximum of 130 amps at 30°C).

An alternative approach, and one that is much faster, is to correct the ampacity of the circuit instead of the ampacity of the conductor by **dividing the ampacity of the circuit by the ambient temperature correction factor.**

For example:

The ampacity of the circuit at 110°F ambient is 125 amps.

The ambient temperature correction factor for 30°C (86°F) ambient conductors is 0.82.

The ampacity of the circuit corrected to 30ºC ambient is 125 amps ÷ 0.82 = 152 amps.

According to NEC® Table 310.15(B)(16), **2/0 AWG** is the **minimum** allowable THWN copper wire size using the 30ºC ampacity value for 152 amps (2/0 AWG THWN copper is rated for a maximum of 175 amps at 30ºC).

This means that at 110ºF ambient, once again we find that 2/0 AWG THWN copper or larger is permissible according to code.

Careful! *Conductor ampacity rating in the National Electrical Code®, like those given in NEC® Table 310.15(B)(16), is the maximum amount of current the conductor is allowed to carry.*

Conductors can always be safely increased in size (for example, from 2/0 AWG to to 3/0 AWG in this problem) but they are generally ***not*** *allowed to be downsized (for example, from 2/0 AWG to 1/0 AWG in this problem).*

When sizing a conductor, the smallest (minimum sized) conductor you may use is the conductor that has an ampacity rating ***equal to or greater than*** *the calculated ampacity of the load after any adjustment factors.*

Even if the conductor has an ampacity rating of one amp less than the ampacity of the circuit, you must use the next larger sized conductor. When using correction factors or other calculations, always save your rounding for the very end and round up to the nearest whole amp for a decimal of 0.5 or greater, and round down for a decimal 0.4 or less.

The answer is: **D and E.**

(Multiple correct AIT question) - Ch. 3.1 National Electrical Code (NEC®)

Problem #73 Solution

This is an **adjustment factor for more than three current-carrying conductors** problem.

TW is a **t**hermoplastic **w**ater-resistant insulated wire for general-purpose wiring.

Since the circuit is rated less than 2,000 volts, made up of insulated conductors, and run in raceway, we will be using NEC® Table 310.15(B)(16) again. For a description of this table, please refer to the previous solution.

The conductor ampacity ratings in NEC® Table 310.15(B)(16) are valid for **not more than (a maximum of) three current-carrying conductors**.

According to NEC® 310.15(B)(3), when there are more than three-current carrying conductors in raceway, the rated conductor ampacity must be reduced by **multiplying** by the percent of value adjustment factor from NEC® Table 310.15(B)(3)(a).

According to the problem, there are already six conductors in the rigid metal conduit for a **total of eight conductors** including the two TW aluminum conductors. The percent of value adjustment factor to reduce the rated ampacity for 8 total conductors in the same raceway from NEC® Table 310.15(B)(3)(a) is **70 (percent)**.

Let's evaluate each of the possible answer choices using the following table. Be sure to use the aluminum ampacities (not copper) for TW on the right side of NEC® Table 310.15(B)(16).

TW Aluminum			
Wire Size	**Ampacity from NEC® Table 310.15(B)(16)**	**Percent of value from NEC® Table 310.15(B)(3)(a)**	**Corrected Ampacity for a total of 8 conductors in raceway**
8 AWG	35	70	25
6 AWG	40	70	28
4 AWG	55	70	39
3 AWG	65	70	46
2 AWG	75	70	53

Notice that the **minimum allowable** TW aluminum wire size for the 50 amp load in the conditions described in the problem is **2 AWG** (2 AWG TW aluminum is rated for a maximum of 53 amps when corrected for 8 conductors in the same raceway).

Out of the possible TW aluminum wire sizes given in the problem, only 2 AWG is allowable according to code.

To compare, if the only conductors in the rigid metal conduit (RMC) were the two TW aluminum conductors for the 50 amp load, then we would **not** need to adjust the ampacity for more than three current-carrying conductors and the minimum TW aluminum wire size required would instead be 4 AWG (4 AWG TW aluminum is rated for a maximum of 55 amps in raceway for not more than three current-carrying conductors).

An alternative approach, and one that is much faster, is to correct the ampacity of the circuit instead of the ampacity of the conductor by **dividing the ampacity of the circuit by percent of value adjustment factor for more than three current-carrying conductors.**

For example:

> The ampacity of the circuit with 8 total current-carrying conductors in the same raceway is 50 amps.
>
> The percent of value adjustment factor for 8 total current-carrying conductors in the same raceway is 70 (percent).
>
> The ampacity of the circuit corrected to not more than three current-carrying conductors in the same raceway is 50 amps ÷ 70% = 71 amps.
>
> According to NEC® Table 310.15(B)(16), **2 AWG** is the **minimum** allowable TW aluminum wire size (2 AWG TW aluminum is rated for a maximum of 75 amps for not more than three current-carrying conductors in the same raceway).

This means that for a total of 8 current-carrying conductors in the same raceway, once again we find that 2 AWG is the minimum allowable TW aluminum wire size according to code.

A wire size of 2 AWG TW aluminum or larger is permissible, but 2 AWG TW aluminum is the largest wire size in the given possible choices.

The answer is: **E.**

(Multiple correct AIT question) - Ch. 3.1 National Electrical Code (NEC®)

Problem #74 Solution

Each possible answer choice is related to overcurrent protection of conductors.

Let's evaluate each possible choice to determine which ones are permissible according to the *2017 National Electrical Code®*:

(A) The maximum standard overcurrent device rating for a branch circuit supplying a 208 volt 175 amp continuous load fed from the minimum required wire size of not more than three 75ºC rated copper current-carrying conductors in raceway is 250 amps.

NEC® 210.19(A)(1)(a) Branch Circuits Not More Than 600 Volts - Branch circuit **continuous loads** must be fed from a conductor with a minimum ampacity of 125% of the continuous load:

175 A(125%) = 219 A.

NEC® Table 310.15(B)(16) - Conductor ampacity for not more than three current-carrying conductors in raceway up to 2,000 volts:

4/0 AWG is the minimum conductor size at 75ºC copper permitted to carry 219 amps with an allowable ampacity of **230 amps**.

230 amps does not correspond with a standard overcurrent device rating listed in NEC® Table 240.6(A).

NEC® 240.4(B) Overcurrent Devices Rated 800 Amperes or Less - This section permits you to select the next higher standard overcurrent device rating from NEC® Table 240.6(A) as long as all three of the following conditions are met:

1. The conductor is not a branch circuit supplying more than one receptacle.

2. The ampacity of the conductor does not match one of the standard overcurrent amp ratings.

3. The next higher standard overcurrent device rating selected does not exceed 800 amps.

The next higher standard overcurrent device rating from NEC® Table 240.6(A) for a conductor with an allowable ampacity of 230 amps is **250 amps**.

250 amps is the **maximum** standard overcurrent device rating that may be used to protect this circuit.

True! **(A) The maximum standard overcurrent device rating for a branch circuit supplying a 208 volt 175 amp continuous load fed from not more than three 75ºC rated copper current-carrying conductors in raceway is 250 amps.**

(B) 225 amps is the maximum standard overcurrent device rating that may be used to protect a conductor with an allowable ampacity of 200 amps.

Since the allowable ampacity of the circuit conductors is 200 amps, and 200 amps is (correspond with) a standard overcurrent protection device rating listed in NEC® Table 240.6(A), we are NOT permitted to select the next higher standard overcurrent device rating according to NEC® 240.(B)(2).

Since we are **not** permitted to round up, the maximum rating of a standard overcurrent device that may be used is 200 amps.

Note: An example of a circuit with an allowable conductor ampacity of 200 amps is one set of 3/0 AWG copper conductors rated for 75ºC for a circuit rated up to 2,000 volts with not more than three current-carrying conductors in conduit according to NEC® Table 310.15(B)(16).

False! **(B) 225 amps is NOT the maximum standard overcurrent device rating that may be used to protect a conductor with an allowable ampacity of 200 amps.**

(C) A three-phase 480 volt panel is fed from two THWN copper 700 kcmil conductors per phase. Each set of three current-carrying conductors is run in separate conduit. 1,000 amps is the maximum standard overcurrent device rating that may be used to protect this circuit.

Since each set of three current-carrying conductors is run in separate conduit, we do not need to apply an ampacity adjustment factor like we would if there were more than three-current carrying conductors, according to NEC® Table 310.15(B)(3)(a).

NEC® Table 310.15(B)(16) - Conductor ampacity for not more than three current-carrying conductors in raceway up to 2,000 volts:

THWN copper 700 kcmil has a 460 amp allowable ampacity.

Two parallel sets (two conductors per phase) of THWN copper 700 kcmil have an allowable ampacity of:

2(460 amps) = 920 amps.

NEC® 240.4(C) Overcurrent Devices Rated over 800 Amperes - For overcurrent devices rated more than 800 amps (for example, 801 amps or more), the allowable ampacity of the conductors must be **equal to or greater** than the rating of the overcurrent device.

This means the overcurrent device rating must be **equal to or smaller** than the allowable ampacity rating of the conductors.

While the NEC® permits selecting the next standard overcurrent device rating for overcurrent devices rated 800 amps or less, it does NOT permit selecting the next standard overcurrent device rating for overcurrent devices rated more than 800 amps.

Since the 920 amp allowable ampacity of the circuit does not correspond with a standard overcurrent device rating listed in NEC® Table 240.6(A) and we are NOT permitted to select the next higher standard overcurrent device rating, we must instead select the next **lower** standard overcurrent device rating to obtain the maximum overcurrent device rating size.

The next lower standard overcurrent device rating from NEC® Table 240.6(A) is **800 amps.**

800 amps is the **maximum** standard overcurrent device rating that may be used to protect this circuit.

False! **(C) A three-phase 480 volt panel is fed from two THWN copper 700 kcmil conductors per phase. Each set of three current-carrying conductors is run in separate conduit. 1,000 amps is NOT the maximum standard overcurrent device rating that may be used to protect this circuit.**

(D) A three-phase 480 volt panel is fed from two THWN copper 700 kcmil conductors per phase. Each set of three current-carrying conductors is run in separate conduit. An inverse time circuit breaker rated for 900 amps supplied by a manufacturer may be used to protect this circuit.

This is the same circuit from the previous choice (C) that we just evaluated. We determined that the conductor allowable ampacity of the circuit is 920 amps.

NEC® 240.6(A) Fuses and Fixed-Trip Circuit Breakers - permits the use of fuses and inverse time circuit breakers with nonstandard ampere ratings that are not listed in NEC® Table 240.6(A).

The standard overcurrent device ratings listed in NEC® Table 240.6(A) are the most common device ratings supplied by just about all electrical manufacturers. However, some manufacturers supply additional ratings for increased flexibility and coordination such as a

900 amp breaker.

NEC® 240.4(C) Overcurrent Devices Rated over 800 Amperes - For overcurrent devices rated more than 800 amps (for example, 801 amps or more), the ampacity of the conductors must be equal to or greater than the rating of the overcurrent device. This means the overcurrent device rating must be equal to or smaller than the ampacity rating of the conductors.

A nonstandard sized 900 amp circuit breaker meets the requirements of NEC® 240.4(C) since the allowable ampacity of the circuit conductors (920 amps) is greater than the 900 amp rating of the circuit breaker.

A nonstandard sized 900 amp inverse time circuit breaker supplied by a manufacturer may be used to protect this circuit.

True! **(D) A three-phase 480 volt panel is fed from two THWN copper 700 kcmil conductors per phase. Each set of three current-carrying conductors is run in separate conduit. An inverse time circuit breaker rated for 900 amps supplied by a manufacturer may be used to protect this circuit.**

(E) The smallest copper-clad aluminum 75°C rated wire size that can be protected by an 80 ampere rated overcurrent device is 3 AWG for a circuit rated up to 2,000 volts with not more than three current-carrying conductors in conduit.

According to NEC® Table 310.15(B)(16), the allowable ampacity of size 3 AWG copper-clad aluminum 75°C rated wire for not more than three current-carrying conductors in conduit for a circuit up to 2,000 volts is 75 amps.

Since 75 amps does not correspond with a standard overcurrent device rating listed in NEC® Table 240.6(A), we are permitted to select the next higher standard overcurrent device rating according to NEC® 240.4(B).

The next higher standard overcurrent device rating listed in NEC® Table 240.6(A) is 80 amps. The 80 ampere rated overcurrent device in the problem is permissible according to code.

Now let's verify if 3 AWG is the smallest size copper-clad aluminum 75°C rated wire that this overcurrent device can protect.

The next smallest copper-clad aluminum 75°C rated wire size according to NEC® Table 310.15(B)(16) is 4 AWG with a 65 amp allowable ampacity.

Since 65 amps does not correspond with a standard overcurrent device rating listed in NEC® Table 240.6(A), we are permitted to select the next higher standard overcurrent device rating according to NEC® 240.4(B).

The next higher standard overcurrent device rating listed in NEC® Table 240.6(A) is 70 amps.

Since 70 amps is the highest allowable overcurrent rating that may be used protect the next smaller copper-clad aluminum 75ºC rated wire size of 4 AWG, the 3 AWG wire size in the problem is indeed the smallest copper-clad aluminum 75ºC rated wire size that can be protected by the 80 ampere rated overcurrent device.

True! **(E) The smallest copper-clad aluminum 75ºC rated wire size that can be protected by an 80 ampere rated overcurrent device is 3 AWG for a circuit rated up to 2,000 volts with not more than three current-carrying conductors in conduit.**

The answer is: **A, D, and E.**

(Multiple correct AIT question) - Ch. 3.1 National Electrical Code (NEC®)

Problem #75 Solution

Most modern residential air conditioning equipment are split systems with an indoor air handler featuring an evaporator coil and evaporator fan motor, and an outdoor condenser featuring a hermetically sealed motor-compressor (commonly known as just a "compressor") and a condenser fan motor.

> *The National Electrical Code®* defines hermetically sealed in NEC® Article 100 - Definitions as: Equipment sealed against the entrance of an external atmosphere where the seal is made by fusion, for example, soldering, brazing, welding, or the fusion of glass to metal.

This problem asks to evaluate codes applicable to the condenser of a split air conditioning system:

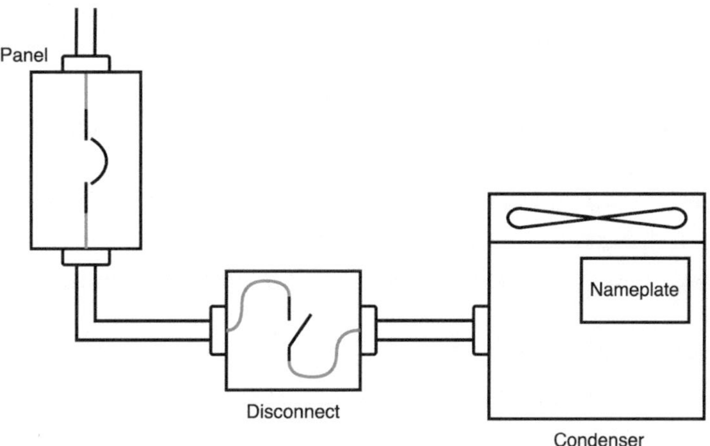

Condenser nameplate:

SUITABLE FOR OUTDOOR USE		
POWER SUPPLY:		
1 PH	60 HZ	208-230 VAC
COMPRESSOR:		
13.50 RLA	72.5 LRA	60 HZ
FAN MOTOR:		
1/4 HP	1.40 FLA	60 HZ
18.3 MINIMUM CIRCUIT AMPS		
30 A MAX FUSE/CKT BKR		
5 KA RMS SYMMETRICAL MAX SHORT CIRCUIT		

Rules for air conditioning and refrigeration equipment can be found in Article 440 of the *2017 National Electrical Code®*. Most sizing calculations for air conditioning equipment can be done very quickly using the nameplate information. In the following solution, we've added additional alternatives in case the complete nameplate information is not provided to you.

Let's evaluate each possible choice to determine which ones are permissible according to the *2017 National Electrical Code®*:

(A) The minimum 60°C copper wire size for this multimotor equipment is 14 AWG.

An HVAC condenser with both a motor-compressor and a condenser fan motor is considered **multimotor combination-load** equipment.

According to NEC® 440.4(B) Multimotor and Combination-Load Equipment, the minimum supply-circuit conductor ampacity shall be provided by the equipment **nameplate**. This value is known as the **branch-circuit selection current** (see NEC® 440.2 Definitions).

The minimum circuit ampacity given in the nameplate in the problem is 18.3 amps.

According to NEC® 440.35 Multimotor and Combination-Load Equipment, the ampacity of the conductors supplying multimotor and combination-load equipment **shall not be less than** the minimum circuit ampacity marked on the equipment in accordance with 440.4(B).

According to NEC® Table 310.15(B)(16), the minimum size 60°C copper wire that can carry 18.3 amps is 12 AWG, which has a maximum allowable ampacity of 20 amperes:

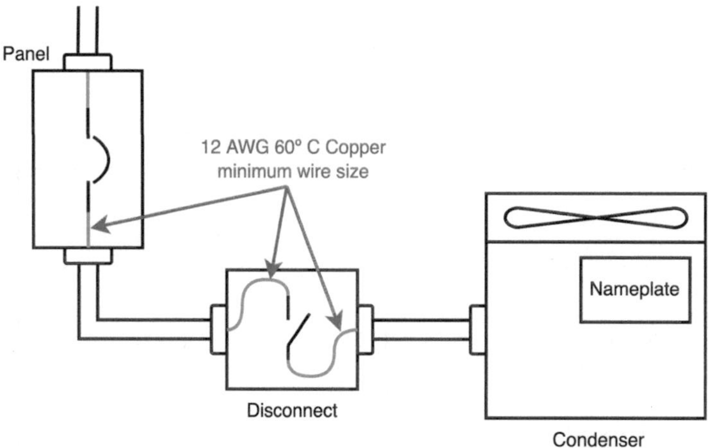

Note: since this is outdoor equipment, we can assume that the wires are protected in a raceway. Since the problem did not mention if there are more than three-current carrying conductors, we can assume we do not need to apply a correction factor. Since the problem

did not specify a hotter than normal ambient temperature we can assume 30ºC, which is typical. For these reasons, we are using NEC® Table 310.15(B)(16) to size the conductor.

What if we did not have the complete nameplate information?

We can arrive at the same value for the branch-circuit selection current as the manufacturer by following the *National Electrical Code®*.

According to NEC® 440.33 Motor-Compressor(s) With or Without Additional Motor Loads, conductors supplying one or more motor-compressor(s) with or without an additional motor load(s) shall have an ampacity not less than the sum of each of the following:

(1) The sum of the rated-load or branch-circuit selection current, whichever is greater, of all motor-compressor(s).
(2) The sum of the full-load current rating of all other motors.
(3) 25 percent of the highest motor-compressor or motor full-load current in the group.

The minimum ampacity of the circuit is therefore:

13.50 A + 1.40 A + 25%(13.50) = 18.3 Amps.

Again, using this information, the minimum size 60ºC copper wire according to NEC® Table 310.15(B)(16) that can carry 18.3 amps is **12 AWG,** which has a maximum allowable ampacity of 20 amperes.

False! (A) The minimum 60ºC copper wire size for this multimotor equipment is NOT 14 AWG, it is 12 AWG.

(B) A 25 amp circuit breaker is permitted for overcurrent protection.

According to NEC® 440.4(B) Multimotor and Combination-Load Equipment, the **maximum** rating of the branch circuit short-circuit and ground-fault protective device shall be provided by the equipment nameplate.

The maximum overcurrent protection rating is given in the nameplate as 30 amps:

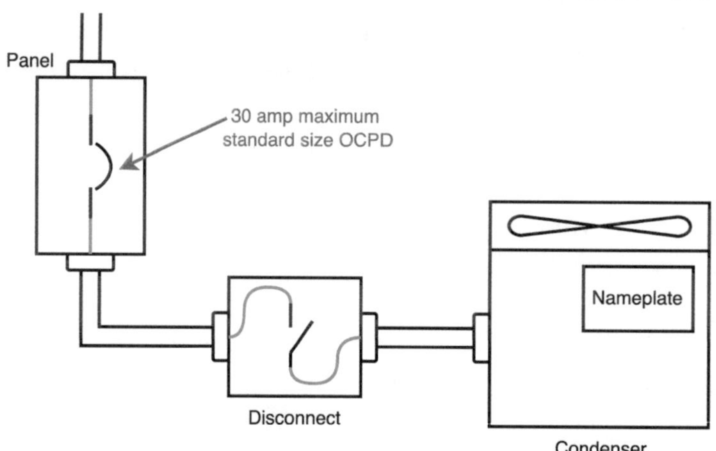

Since this value represents the **maximum** overcurrent device rating, the option of 25 amps given in this answer choice is still permitted as far as the *National Electrical Code®* is concerned since it is **less than** the maximum rating.

What if we did not have the complete nameplate information?

We can arrive at the same value for the maximum standard sized overcurrent protection device (OCPD) as the manufacturer by following the *National Electrical Code®*.

According to NEC® 440.22(B) Rating or Setting for Equipment, Where the equipment incorporates more than one hermetic refrigerant motor-compressor or a hermetic refrigerant motor-compressor and other motors or other loads, the equipment short-circuit and ground-fault protection shall comply with 430.53 and 440.22(B)(1) and (B)(2).

NEC® 430.53 just specifies the conditions when several motors or loads can be powered by the same branch circuit, which the multimotor combination load condenser satisfies according to NEC® 430.53(C).

NEC® 440.22(B)(1) is used when the motor-compressor is the largest load. According to the nameplate, the 13.50 rated-load amp (RLA) motor-compressor is larger (draws more current) than the 1.40 full-load amps (FLA) fan motor. NEC® 440.22(B)(1) instructs that the rating of the OCPD shall not exceed the value specified in 440.22(A) for the motor-compressor plus the sum of the ratings of the other loads supplied.

NEC® 440.22(A) Rating or Setting for Individual Motor-Compressor, instructs that the OCPD shall not exceed 225% of the motor rated load current (RLA).

We can skip 440.22(B)(2) since it only applies when the motor-compressor is **not** the largest load.

In other words, the maximum OCPD rating cannot exceed the sum of 225% of the motor-compressor rated load current (RLA) and the rating of the condenser fan motor. The **maximum** OCPD rating is:

225%(13.50 A) + 1.40 A = 31.8 amps.

31.8 amps does not correspond with a standard overcurrent device rating listed in NEC® Table 240.6(A) so we must round down to the next NEC® 240.6(A) standard overcurrent device rating of **30 amps**, just like what the manufacturer provided on the nameplate.

*Careful! The code rule that permits the next higher overcurrent device rating to be selected when the maximum calculated rating does not correspond to a standard overcurrent device rating for **conductors** per NEC® 240.4(B) does **not** apply here. NEC® 240.4(G) Overcurrent Protection for Specific Conductor Applications points to Article 440, Parts III and VI for air conditioning and refrigeration. There are no provisions, exceptions, or rules in Article 440, Parts III and VI that permit selecting the next side overcurrent device rating when the calculated rating does not correspond with a higher size.*

Again, since this value represents the **maximum** overcurrent device rating, the option of 25 amps given in this answer choice is still permitted as far as the *National Electrical Code®* is concerned since it is **less than** the maximum rating.

*Note: typically, when sizing conductors and overcurrent protection for motors, according to NEC® 430.6(A)(1), the National Electrical Code® instructs us **NOT** to use the nameplate full-load amps (FLA) and to use the NEC® motor full-load current (FLC) tables based on voltage, horsepower, and phase instead (see NEC® Tables 430.248–250). However, this does **not** apply specifically for air conditioning motors such as motor-compressors and condenser fan or evaporator blower motors. See NEC® 440.6(A) for a single motor-compressor and 440.6(B) for multimotor equipment (like a condenser made up of a motor-compressor and a fan motor). Note the NEC®'s use of "rated-load current marked on the nameplate" and "full-load current marked on the nameplate".*

True! **(B) A 25 amp circuit breaker is permitted for overcurrent protection since it is less than the rated maximum.**

(C) The minimum full-load current equivalent disconnect rating permitted rounded to the nearest ampere is 17.

We can determine the **minimum** full-load current equivalent disconnect rating for a combination load. According to NEC® 440.12(B)(2) Full-Load Current Equivalent, the ampere rating of the disconnecting means shall be at least 115 percent of the sum of all currents at the rated-load condition determined in accordance with 440.12(B)(1).

For motors **other than** the motor-compressors and fan or blower motors, NEC® Tables 430.248–250 are used to determine the full-load current (FLC). For motor-compressors and fan or blower motors we again use the equipment nameplate according to NEC® 440.6(B).

Since all we have in this example is one standard condenser from a split air conditioning system consisting of one motor-compressor and one condenser fan motor, all we need to do is sum their rated currents from the nameplate and multiply by 115% to determine the **minimum** disconnect rating in amps:

115%(13.50 A + 1.40 A) = 17.1 A

17 amps is the **minimum** disconnect rating rounded to the nearest ampere.

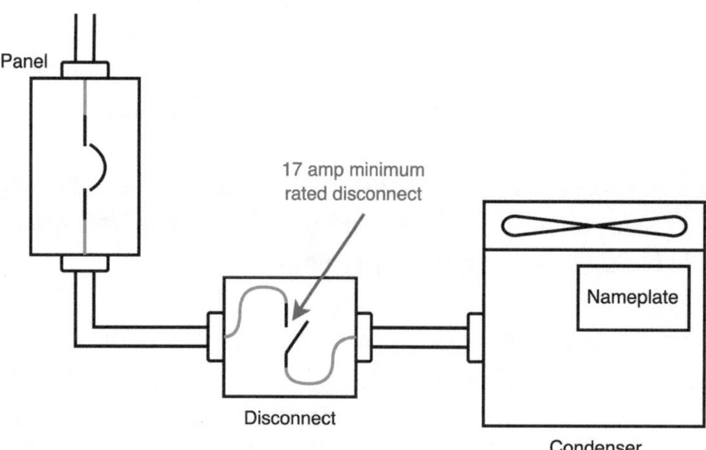

*Note: since we are **not** being asked to size the disconnect based on horsepower (instead of amps), we can ignore the fact that the nameplate branch-circuit selection current (18.3 minimum circuit amps) is greater than 115% of the sum of all currents at rated-load conditions. If we were selecting a horsepower-rated disconnect, this would apply per NEC® 440.12.(B)(1)(a).*

True! **(C) The minimum full-load current equivalent disconnect rating permitted rounded to the nearest ampere is 17.**

(D) The disconnect for this condenser must be within 30 feet of the equipment.

Article 440 makes no mention of the specified amount of feet that the disconnect must be installed within, only that it must be within eyesight and readily accessible:

NEC® 440.14 Location. Disconnecting means shall be located within sight from, and readily accessible from the air conditioning or refrigerating equipment.

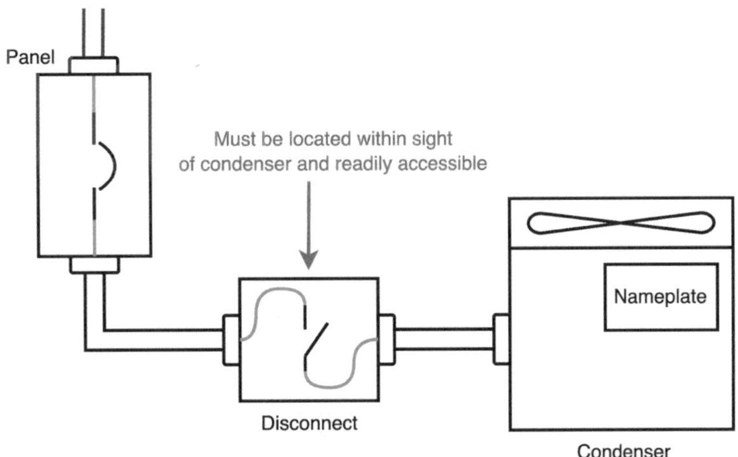

Note: if a circuit breaker is used for the overcurrent protection device (OCPD) and is located within sight and readily accessible from the equipment, then it may also be used as the disconnect.

False! **(D) The disconnect for this condenser does NOT need to be within 30 feet of the equipment, it must be within sight from and readily accessible from the equipment.**

(E) A motor controller with a 10 kA short-circuit current rating (SCCR) is permitted.

The rules regarding short-circuit current rating (SCCR) for the motor controller are:

NEC® 440.10(A) Installation. Motor controllers of multimotor and combination-load equipment shall not be installed where the available short-circuit current exceeds its short-circuit current rating.

Or more simply stated, the short-circuit current rating (SCCR) of the motor controller must be equal to or greater than the available short-circuit current of the equipment.

According to the condenser nameplate, the available short-circuit current is 5 kA. Since the motor controller has a short-circuit current rating of 10 kA, it is indeed permissible to be used with this equipment.

True! **(E) A motor controller with a 10 kA short-circuit current rating (SCCR) is**

permitted.

The answer is: **B, C, and E.**

(Multiple correct AIT question) - Ch. 3.1 National Electrical Code (NEC®)

Problem #76 Solution

Cable tray is a structural support and cable management system to help route conductors and cables. Cable tray is mostly used in maintenance heavy locations since conductors and cables can be installed or removed rather than pulling wire in conduit:

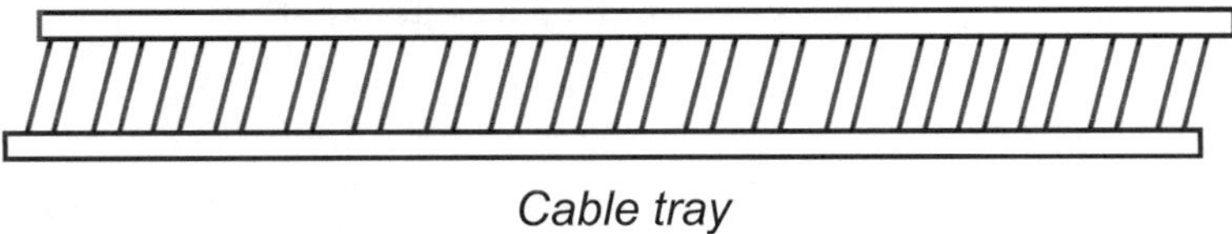

Cable tray

The *National Electrical Code®* defines the rules for cable tray in Article 392 as:

NEC® 392.2 Definition. Cable Tray System. A unit or assembly of units or sections and associated fittings forming a structural system used to securely fasten or support cables and raceways.

Let's evaluate each possible statement to identify which are true and permitted for cable tray installations according to the *2017 National Electrical Code®*:

(A) Cable tray is considered as a raceway when determining conductor allowable ampacity ratings for single conductors.

Cable tray is **not** a type of raceway, it is a structural support system.

The rules for determining the minimum allowable ampacity of single conductors in cable tray are defined in NEC® 392.80(A)(2) Single-Conductor Cables. The rules mostly depend on the size of the single conductor, the spacing in the tray, and if the tray is covered or not.

For example, NEC® 392.80(A)(2)(c) permits the allowable ampacity of single conductors 1/0 AWG and larger in uncovered tray with a minimum spacing of one cable diameter of air between them to be determined according to NEC® Table 310.15(B)(17) and NEC® Table 310.15(B)(19) which are **free air** (not raceway) conductor allowable ampacities tables.

False! (A) Cable tray is NOT NECESSARILY considered as a raceway when determining conductor allowable ampacity ratings for single conductors.

(B) Cable tray is only permitted to be installed in industrial locations.

According to NEC® 392.10 Use Permitted, this is **not** true: cable tray installations **shall not** be limited to industrial establishments.

False! **(B) Cable tray is NOT only permitted to be installed in industrial locations.**

(C) Parallel circuits must be installed so that only one conductor per phase is located in the same cable tray.

According to NEC® 392.20(C) Connected in Parallel, this is **not** true.

While each set of a parallel circuit must be grouped together with not more than one conductor per phase, neutral, or ground, it does **not** mention anything about each parallel circuit being installed in separate cable trays.

For example, a two conductor per-phase parallel circuit would **not** be permitted to be installed in the cable tray in the following order:

A A B B C C N N

Where:

A = one A-phase conductor
B = one B-phase conductor
C = one C-phase conductor
N = one neutral conductor

Instead, they must be grouped in the following order:

A B C N A B C N

False! **(C) Parallel circuits ARE NOT REQUIRED to be installed so that only one conductor per phase is located in the same cable tray.**

(D) Where all cables are multiconductor size 4/0 AWG and larger, the sum of the diameters of all cables may not exceed 50% of ladder-type cable tray width.

According to NEC® 392.22(A)(1)(a) this is **not** true:

NEC® 392.22(A)(1)(a) Where all of the cables are 4/0 AWG or larger, the sum of the diameters of all cables shall not exceed the cable tray width, and the cables shall be installed in a single layer.

False! **(D) For multiconductor size 4/0 AWG cables and larger, the sum of the diameters of all cables may not exceed THE WIDTH of ladder-type cable tray.**

(E) **Metal cable trays are permitted to be used as equipment grounding conductors in locations where continuous maintenance and supervision ensure that only qualified persons service the cable tray systems.**

According to NEC® 392.60(A) this is true:

NEC® 392.60(A) Metal Cable Trays. Metal cable trays **shall be permitted** to be used as equipment grounding conductors where continuous maintenance and supervision ensure that qualified persons service the installed cable tray system.

True! **(E) Metal cable trays are permitted to be used as equipment grounding conductors in locations where continuous maintenance and supervision ensure that only qualified persons service the cable tray systems.**

The answer is: **E.**

(Multiple correct AIT question) - Ch. 3.1 National Electrical Code (NEC®)

Problem #77 Solution

According to the National Electrical Code® in Article 100 - Definitions, a **branch circuit** is the circuit conductors between the final overcurrent device protecting the circuit and the outlet(s).

Receptacle outlets are fed by branch circuits which are covered in Article 210 of the *National Electrical Code®*.

Let's evaluate each possible statement to identify which are true and permitted for receptacles according to the *2017 National Electrical Code®*:

(A) The rating of a branch circuit is determined by the allowable ampacity rating of the branch-circuit conductor.

This is **false** according to NEC® 210.18 Rating. Branch circuits shall be rated in accordance with the maximum permitted ampere rating or setting of the overcurrent device. The rating for other than individual branch circuits shall be 15, 20, 30, 40, and 50 amperes. Where conductors of higher ampacity are used for any reason, the ampere rating or setting of the specified overcurrent device shall determine the circuit rating.

The rating of the branch-circuit overcurrent device, not the minimum allowable ampacity of the conductor, determines the rating of the branch circuit.

False! **(A) The rating of a branch circuit is NOT determined by the allowable ampacity rating of the branch-circuit conductor.**

(B) A single receptacle installed on an individual branch circuit must have an equal or greater ampere rating than the overcurrent protection device of the circuit.

This is **true** according to NEC® 210.21(B)(1) Single Receptacle on an Individual Branch Circuit. A single receptacle installed on an individual branch circuit shall have an ampere rating not less than that of the branch circuit.

A single receptacle installed on an individual branch circuit means there is one receptacle and no other load on the circuit. Common examples would be a single receptacle installed on its own individual branch circuit to supply power to a kitchen stove or clothes dryer.

For example, a single 30 ampere rated receptacle on an individual branch circuit must be protected by an overcurrent device rated at least 30 amps or less.

True! **(B) A single receptacle installed on an individual branch circuit must have an equal or greater ampere rating than the overcurrent protection device of the circuit.**

(C) A branch circuit protected by a 20 ampere rated overcurrent protection device may supply multiple receptacles rated for both 15 amperes and 20 amperes.

This is **true** according to NEC® 210.21(B)(3) Receptacle Ratings. Where connected to a branch circuit supplying two or more receptacles or outlets, receptacle ratings shall conform to the values listed in Table 210.21(B)(3), or, where rated higher than 50 amperes, the receptacle rating shall not be less than the branch-circuit rating.

NEC® Table 210.21(B)(3) Receptacle Ratings for Various Size Circuits permits both 15 ampere and 20 ampere rated receptacles to be supplied by the same 20 ampere rated circuit.

Compared to a single receptacle installed on an individual branch circuit, different rules apply when a branch circuit supplies more than one receptacle.

True! **(C) A branch circuit protected by a 20 ampere rated overcurrent protection device may supply multiple receptacles rated for both 15 amperes and 20 amperes.**

(D) 15 amperes is the maximum rating for a cord-and-plug equipment powered by a receptacle on a 20 ampere rated branch circuit.

According to NEC® 210.23(A)(1) Cord-and-Plug-Connected Equipment Not Fastened in Place. The rating of any one cord-and-plug-connected utilization equipment not fastened in place **shall not exceed 80 percent** of the branch-circuit ampere rating.

Cord-and-plug equipment refers to any flexible cord and attachment plug equipment such as a vacuum cleaner, coffee machine, or generally any equipment you would power by plugging into a receptacle.

The maximum rating of a cord-and-plug equipment supplied by a 20 ampere rated branch circuit is 16 amperes, **not** 15 amperes:

80%(20 A) = 16 A.

False! **(D) 15 amperes is NOT the maximum rating for a cord-and-plug equipment powered by a receptacle on a 20 ampere rated branch circuit.**

(E) Branch circuits in a multiple occupancy dwelling unit may not supply loads in an adjacent dwelling unit.

This is true according to NEC® 210.25(A) Dwelling Unit Branch Circuits. Branch circuits in each dwelling unit shall supply only loads within that dwelling unit or loads associated only with that dwelling unit.

True! **(E) Branch circuits in a multiple occupancy dwelling unit may not supply loads in an adjacent dwelling unit.**

The answer is: **B, C, and E.**

(Multiple correct AIT question) - Ch. 3.1 National Electrical Code (NEC®)

Problem #78 Solution

A feeder is generally any conductor between the initial service conductor and the final branch-circuit conductor.

According to the National Electrical Code® in Article 100 - Definitions, a **feeder** is any circuit conductor between the service equipment, the source of a separately derived system, or other power supply source and the final branch-circuit overcurrent device.

Feeders are covered in Article 215 of the *National Electrical Code®*.

Let's evaluate each possible statement to identify which are true and permitted for feeder conductors according to the *2017 National Electrical Code®*:

(A) If adjustment or correction factors are not required for a feeder up to 600 volts, then the allowable ampacity must be equal to or greater than the sum of 125% of the continuous loads and 100% of the noncontinuous loads.

This is **true** according to NEC® 215.2(A)(1)(a): Where a feeder supplies continuous loads or any combination of continuous and noncontinuous loads, the minimum feeder conductor size shall have an allowable ampacity not less than the noncontinuous load plus 125 percent of the continuous load.

True! **(A) If adjustment or correction factors are not required for a feeder up to 600 volts, then the allowable ampacity must be equal to or greater than the sum of 125% of the continuous loads and 100% of the noncontinuous loads.**

(B) If adjustment or correction factors are required for a feeder up to 600 volts, then the allowable ampacity must be equal to or greater than the maximum load.

This is **true** according to NEC® 215.2(A)(1)(b): The minimum feeder conductor size shall have an allowable ampacity not less than the maximum load to be served after the application of any adjustment or correction factors.

If you have to apply any ampacity correction factors, such as if there are more than three current-carrying conductors in the raceway, or if the ambient temperature is hotter than the ambient temperature rating of the conductor, the NEC® will permit you to instead size the feeder conductor based on the total maximum load being served, which is the sum of 100% continuous load and 100% noncontinuous load, instead of the sum of 125% continuous load and 100% noncontinuous load as we saw before.

True! **(B) If adjustment or correction factors are required for a feeder up to 600 volts,**

then the allowable ampacity must be equal to or greater than the maximum load.

(C) If the neutral conductor of a feeder up to 600 volts is not connected to an overcurrent device, it must have an allowable ampacity of at least the sum of 100% of the noncontinuous load and 100% of the continuous load.

This is **true** according to Exception No. 3 of NEC® 215.2(A)(1)(a): Grounded conductors that are not connected to an overcurrent device shall be permitted to be sized at 100 percent of the continuous and noncontinuous load.

The neutral is a grounded conductor.

For example, consider a three-phase four-wire feeder supplying power to a three-phase panel with a main breaker. The three-phase panel main breaker only has three poles total, one for each of the A, B, and C phase conductors to terminate on. The neutral (grounded conductor) from the feeder would land on the neutral bar.

As long as this condition is met, this exception allows the neutral conductor to be sized at a new minimum of the sum of 100% continuous load and 100% noncontinuous load, instead of the sum of 125% continuous load and 100% noncontinuous load.

True! **(C) If the neutral conductor of a feeder up to 600 volts is not connected to an overcurrent device, it must have an allowable ampacity of at least the sum of 100% of the noncontinuous load and 100% of the continuous load.**

(D) The minimum rating for feeder overcurrent protection is the sum of the continuous loads and noncontinuous loads.

This is **false** according to NEC® 215.3 Overcurrent Protection. Feeders shall be protected against overcurrent in accordance with the provisions of Part I of Article 240. Where a feeder supplies continuous loads or any combination of continuous and noncontinuous loads, the rating of the overcurrent device shall not be less than the noncontinuous load plus 125 percent of the continuous load.

False! **(D) The minimum rating for feeder overcurrent protection is NOT the sum of the continuous loads and noncontinuous loads.**

(E) Up to two sets of 3-wire feeders may share a common neutral conductor.

This is **false** according to NEC® 215.4(A) Feeders with Common Neutral. Up to three sets of 3-wire feeders or two sets of 4-wire or 5-wire feeders shall be permitted to utilize a common neutral.

Each set is a parallel circuit with an additional conductor per each phase. For example, a parallel circuit with up to three conductors per phase (three sets) may share a common neutral for 3-wire feeders.

False! **(E) Up to THREE sets of 3-wire feeders may share a common neutral conductor.**

The answer is: **A, B, and C.**

(Multiple correct AIT question) - Ch. 3.1 National Electrical Code (NEC®)

Problem #79 Solution

Branch-circuit, feeders, and service load calculations are covered in Article 220 of the *National Electrical Code®.*

Unlike **dwelling occupancies,** for which the *National Electrical Code®* specifies the load in volt-amps per square foot for general-purpose branch circuits (NEC® Table 220.12), for **commercial occupancies,** the *National Electrical Code®* instead specifies how much load in volt-amps to account for receptacles for general-purpose branch circuits.

According to NEC® 220.14(I) Receptacle Outlets:

"Except as covered in 220.14(J) and (K), ***receptacle outlets shall be calculated at not less than 180 volt-amperes for each single or for each multiple receptacle on one yoke****. A single piece of equipment consisting of a multiple receptacle comprised of four or more receptacles shall be calculated at not less than 90 volt-amperes per receptacle. This provision shall not be applicable to the receptacle outlets specified in 210.11(C)(1) and (C)(2)."*

If you're not familiar with the term **yoke**, it is just the fastening bracket that receptacles and switches are mounted on that screw into the junction box (note that yoke does not appear in NEC® Article 100 Definitions):

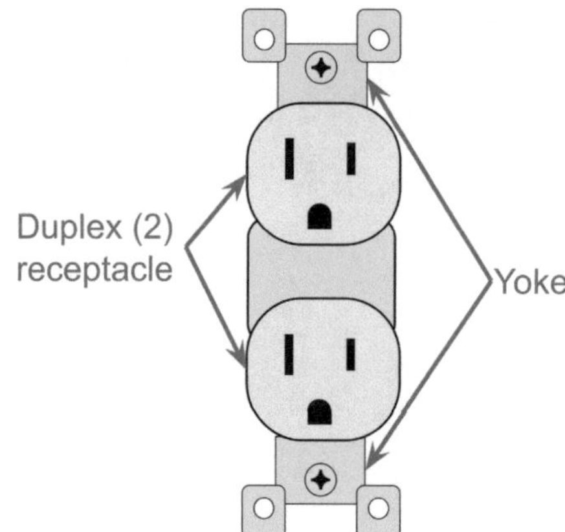

Duplex means two (2), and a duplex receptacle is typically what you imagine when you think of a standard power outlet in your home or office.

First let's determine the maximum load in volt-amps permitted on a 20 amp rated 120 volt circuit using the single-phase power formula:

$$|S_{1\phi}| = |V_P||I_P|$$

$$|S_{1\phi}| = (120\,V)(20\,A)$$

$$|S_{1\phi}| = 2,400\,VA$$

One 120 volt 20 ampere circuit can support **2,400 volt-amps** of load.

Since each duplex receptacle is counted as 180 VA according to NEC® 220.14(I) for commercial occupancies, we can determine the maximum number permitted by dividing the total volt-amps by the volt-amp per receptacle:

$$\frac{2{,}400\,VA}{180\,VA} = 13.3$$

Since we can't have 0.3 of a receptacle, we must round down to 13 receptacles.

A maximum of 13 single-yoke duplex receptacles are permitted on one 20 ampere rated 120 volt general-purpose branch circuit in a commercial occupancy according to the *2017 National Electrical Code®.*

The answer is: **13**

(Multiple correct AIT question) - Ch. 3.1 National Electrical Code (NEC®)

Problem #80 Solution

Grounding and bonding is located in Article 250 of the *National Electrical Code®* and many of the terms are defined in Article 100 Definitions.

There are many helpful diagrams throughout Article 250 in the Handbook version of the *National Electrical Code®* to help illustrate the different grounding techniques and requirements.

1. NEC® Article 100 Definitions. **Bonding Jumper, System**. The connection between the grounded circuit conductor and the supply-side bonding jumper, or the equipment grounding conductor, or both, at a separately derived system.

NEC® 250.30(A)(1) **System Bonding Jumper.** This connection shall be made at any single point on the separately derived system from the source to the first system disconnecting means or overcurrent device, or it shall be made at the source of a separately derived system that has no disconnecting means or overcurrent devices. The system bonding jumper shall remain within the enclosure where it originates.

The purpose of the **system bonding jumper** is to bond the power supply transformer to ground:

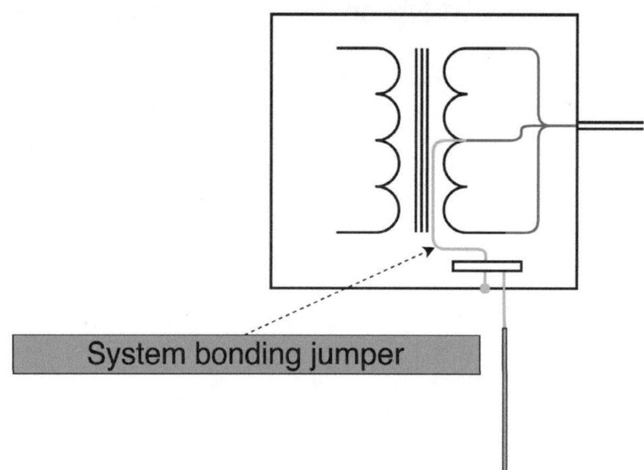

2. NEC® Article 100 Definitions. **Bonding Jumper, Main**. The connection between the grounded circuit conductor and the equipment grounding conductor at the service.

NEC® 250.24(B) **Main Bonding Jumper.** For a grounded system, an unspliced main bonding jumper shall be used to connect the equipment grounding conductor(s) and the service-disconnect enclosure to the grounded conductor within the enclosure for each service disconnect in accordance with 250.28.

The purpose of the **main bonding jumper** is to bond the neutral conductor on the load side of the service to ground. Here, the main breaker in the panel is acting as the service disconnect. If there was a separate disconnect upstream between the panel and the transformer, then the main bonding jumper would be located in that enclosure instead.

It is important to note that there is only one main bonding conductor per service disconnect:

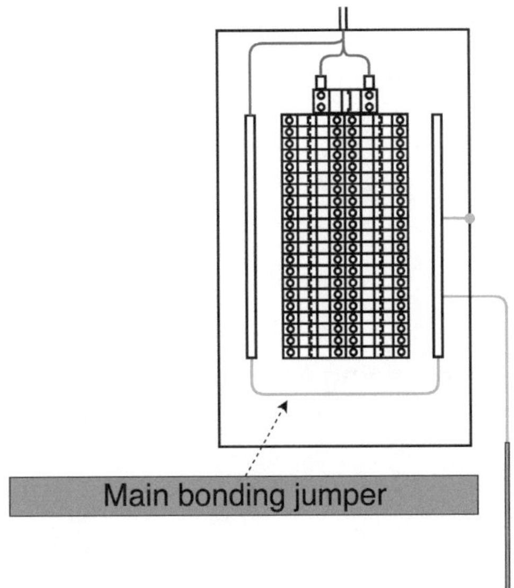

3. NEC® Article 100 Definitions. **Grounding Conductor, Equipment (EGC).** The conductive path(s) that provides a ground-fault current path and connects normally non-current-carrying metal parts of equipment together and to the system grounded conductor or to the grounding electrode conductor, or both.

NEC® 250.86 **Other Conductor Enclosures and Raceways**. Metal enclosures and raceways for other than service conductors shall be connected to the equipment grounding conductor.

The purpose of the **equipment grounding conductor** is to ground metal equipment that is not intended to be energized (such as metal junction boxes, metal panels, metal lighting fixtures, metal conduit, etc) in order to provide a path for ground-fault current to lead back to the power source.

Metal conduit is often used as the equipment grounding conductor between metal enclosures. In the diagram in the problem, the metal panel enclosure is grounded with an **equipment grounding conductor.** The metal enclosure of the transformer is also grounded with an **equipment grounding conductor;** however, it was not a possible answer choice:

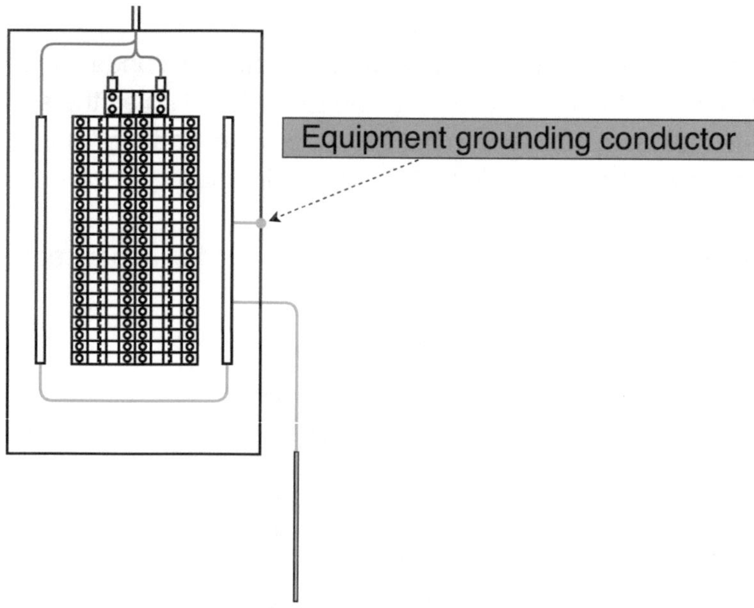

4. NEC® Article 100 Definitions. **Grounding Electrode Conductor.** A conductor used to connect the system grounded conductor or the equipment to a grounding electrode or to a point on the grounding electrode system.

NEC® 250.24(D) **Grounding Electrode Conductor**. A grounding electrode conductor shall be used to connect the equipment grounding conductors, the service-equipment enclosures, and, where the system is grounded, the grounded service conductor to the grounding electrode(s).

The purpose of the **grounding electrode conductor** is to connect the grounding electrode to ground:

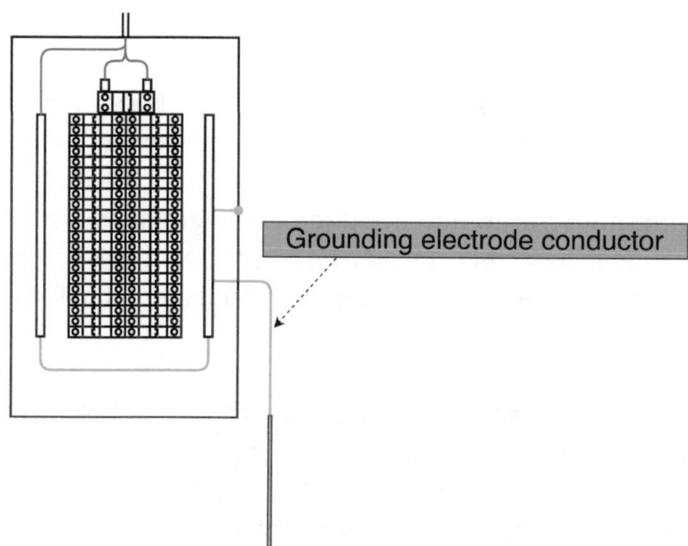

5. NEC® Article 100 Definitions. **Grounding Electrode.** A conducting object through which a direct connection to earth is established.

NEC® 250.50 **Grounding Electrode System**. All grounding electrodes that are present at each building or structure served shall be bonded together to form the grounding electrode system.

A more common name for **grounding electrode** is a grounding rod:

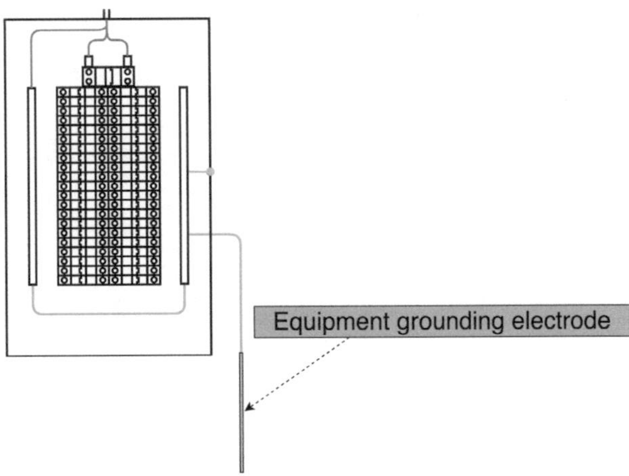

Here is the completed drag and drop diagram:

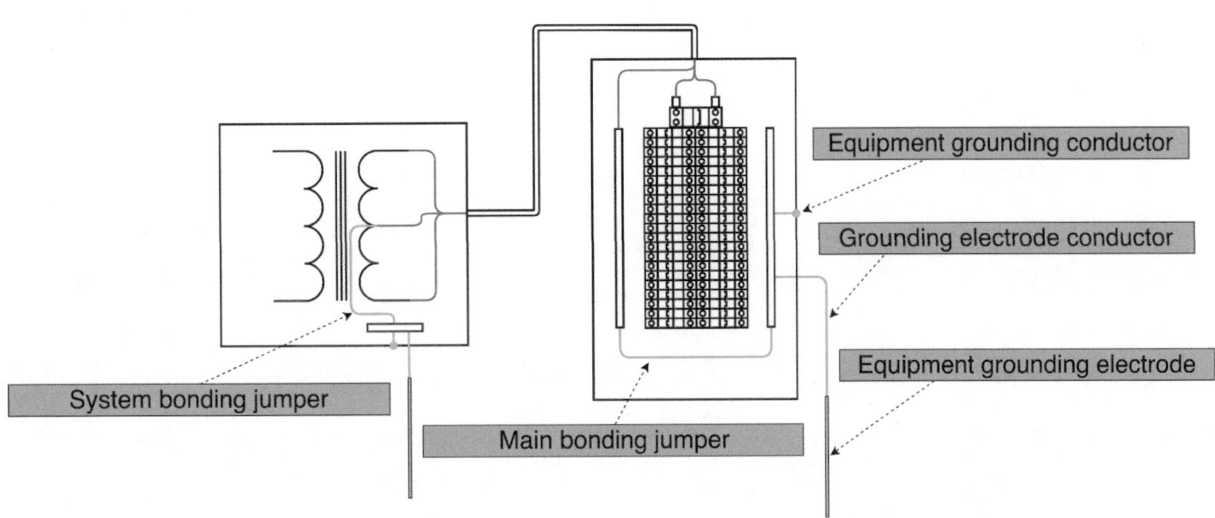

The answer is: **See the completed drag and drop diagram above.**

(Multiple correct AIT question) - Ch. 3.1 National Electrical Code (NEC®)

Answer Key

#	Exam Subject	Solution Page	Answer
1	Measurement and Instrumentation	62	ABCDE
2	Measurement and Instrumentation	63	5
3	Measurement and Instrumentation	65	See Graph
4	Measurement and Instrumentation	66	340
5	Measurement and Instrumentation	68	0.6
6	Measurement and Instrumentation	70	BC
7	Applications	72	51,697
8	Applications	73	BCD
9	Applications	76	46
10	Applications	77	46.6%
11	Applications	78	AD
12	Applications	81	ABDE
13	Applications	84	$3,168
14	Applications	85	BCD
15	Applications	86	0.7
16	Analysis	88	See Diagram
17	Analysis	89	AE
18	Analysis	93	7.2
19	Analysis	95	1.5
20	Analysis	101	See Diagram
21	Analysis	105	0 + j0.36 pu
22	Analysis	107	BCDE
23	Analysis	113	285
24	Analysis	114	0.4
25	Analysis	117	45
26	Devices and Power Electronics	119	3.8
27	Devices and Power Electronics	121	C

Answer Key

#	Exam Subject	Solution Page	Answer
28	Devices and Power Electronics	123	ABDE
29	Devices and Power Electronics	126	CD
30	Devices and Power Electronics	128	See Diagram
31	Devices and Power Electronics	130	See Diagram
32	Induction and Synchronous Machines	133	199
33	Induction and Synchronous Machines	138	19
34	Induction and Synchronous Machines	140	See Diagram
35	Induction and Synchronous Machines	141	1
36	Induction and Synchronous Machines	143	See Diagram
37	Induction and Synchronous Machines	145	See Diagram
38	Induction and Synchronous Machines	147	BCDE
39	Electric Power Devices	150	677
40	Electric Power Devices	153	AC
41	Electric Power Devices	156	See Diagram
42	Electric Power Devices	162	ADE
43	Electric Power Devices	164	BC
44	Electric Power Devices	165	3.5%
45	Electric Power Devices	167	DE
46	Electric Power Devices	168	112
47	Electric Power Devices	170	ABCD
48	Power System Analysis	172	BDE
49	Power System Analysis	178	ACD
50	Power System Analysis	184	ABC
51	Power System Analysis	190	See Diagram
52	Power System Analysis	194	ABD
53	Power System Analysis	202	C
54	Power System Analysis	209	40

Answer Key

#	Exam Subject	Solution Page	Answer
55	Power System Analysis	212	0.26
56	Power System Analysis	216	AE
57	Power System Analysis	220	114
58	Protection	222	ABC
59	Protection	225	See Diagram
60	Protection	228	See Diagram
61	Protection	233	AB
62	Protection	239	B
63	Protection	241	See Diagram
64	Protection	245	BDE
65	Protection	247	ABDE
66	Protection	250	AB
67	Protection	254	See Diagram
68	Protection	256	1,500 amps, 4,000 amps, 8,000 amps, 40,000 amps
69	Protection	258	BC
70	Codes and Standards	263	See Diagram
71	Codes and Standards	267	BC
72	Codes and Standards	270	DE
73	Codes and Standards	274	E
74	Codes and Standards	276	ADE
75	Codes and Standards	281	BCE
76	Codes and Standards	289	E
77	Codes and Standards	292	BCE
78	Codes and Standards	295	ABC
79	Codes and Standards	298	13
80	Codes and Standards	300	See Diagram

Index of Problems by AIT Question Type

#	AIT Type	Solution Page	Answer
30	Point and click	128	See Diagram
31	Point and click	130	See Diagram

#	AIT Type	Solution Page	Answer
3	Drag and drop	65	See Graph
16	Drag and drop	88	See Diagram
20	Drag and drop	101	See Diagram
34	Drag and drop	140	See Diagram
36	Drag and drop	143	See Diagram
37	Drag and drop	145	See Diagram
41	Drag and drop	156	See Diagram
51	Drag and drop	190	See Diagram
59	Drag and drop	225	See Diagram
60	Drag and drop	228	See Diagram
63	Drag and drop	241	See Diagram
67	Drag and drop	254	See Diagram
70	Drag and drop	263	See Diagram
80	Drag and drop	300	See Diagram

Index of Problems by AIT Question Type

#	AIT Type	Solution Page	Answer
2	Fill in the blank	63	5
4	Fill in the blank	66	340
5	Fill in the blank	68	0.6
7	Fill in the blank	72	51,697
9	Fill in the blank	76	46
10	Fill in the blank	77	46.6%
13	Fill in the blank	84	$3,168
15	Fill in the blank	86	0.7
18	Fill in the blank	93	7.2
19	Fill in the blank	95	1.5
21	Fill in the blank	105	0 + j0.36 pu
23	Fill in the blank	113	285
24	Fill in the blank	114	0.4
25	Fill in the blank	117	45
26	Fill in the blank	119	3.8
32	Fill in the blank	133	199
33	Fill in the blank	138	19
35	Fill in the blank	141	1
39	Fill in the blank	150	677
44	Fill in the blank	165	3.5%
46	Fill in the blank	168	112
54	Fill in the blank	209	40
55	Fill in the blank	212	0.26
57	Fill in the blank	220	114
68	Fill in the blank	256	1,500 amps, 4,000 amps, 8,000 amps, and 40,000 amps
79	Fill in the blank	298	13

Index of Problems by AIT Question Type

#	AIT Type	Solution Page	Answer
1	Multiple correct	62	ABCDE
6	Multiple correct	70	BC
8	Multiple correct	73	BCD
11	Multiple correct	78	AD
12	Multiple correct	81	ABDE
14	Multiple correct	85	BCD
17	Multiple correct	89	AE
22	Multiple correct	107	BCDE
27	Multiple correct	121	C
28	Multiple correct	123	ABDE
29	Multiple correct	126	CD
38	Multiple correct	147	BCDE
40	Multiple correct	153	AC
42	Multiple correct	162	ADE
43	Multiple correct	164	BC
45	Multiple correct	167	DE
47	Multiple correct	170	ABCD
48	Multiple correct	172	BDE
49	Multiple correct	178	ACD
50	Multiple correct	184	ABC
52	Multiple correct	194	ABD
53	Multiple correct	202	C
56	Multiple correct	216	AE
58	Multiple correct	222	ABC
61	Multiple correct	233	AB
62	Multiple correct	239	B

Index of Problems by AIT Question Type

#	AIT Type	Solution Page	Answer
64	Multiple correct	245	BDE
65	Multiple correct	247	ABDE
66	Multiple correct	250	AB
69	Multiple correct	258	BC
71	Multiple correct	267	BC
72	Multiple correct	270	DE
73	Multiple correct	274	E
74	Multiple correct	276	ADE
75	Multiple correct	281	BCE
76	Multiple correct	289	E
77	Multiple correct	292	BCE
78	Multiple correct	295	ABC

More by Zach Stone, P.E. and Electrical PE Review, INC.

I hope you enjoyed learning from the unique AIT practice problems and detailed solutions offered in this book. Below are links to more study material available by Zach Stone, P.E. and Electrical PE Review, INC for the *NCEES® Power PE Exam.*

The 2nd edition of our first practice exam, the ***Electrical Engineering PE Practice Exam and Technical Study Guide*** has been updated for the CBT format and is now available on Amazon.com:

Electrical Engineering PE Practice Exam and Technical Study Guide - 2nd Edition Updated for the CBT Format

Amazon.com

★★★★★ **Best Study Guide for taking the CBT Power PE Exam**

Reviewed in the United States on October 21, 2021

Verified Purchase

I passed the CBT Power PE exam on my first attempt. I used several products to review for the PE, but Zach Stone's product really provided problems that were closer to what I encountered on the exam than any other manual I used.

More by Zach Stone, P.E. and Electrical PE Review, INC.

More PE practice exams available on Amazon for the Power PE Exam

Scan the QR code to the right or go to www.amazon.com and search for "*Zach Stone PE*"

More Practice Exams:

Sign up for the 100% free trial of our popular online review course for the *NCEES® Power PE Exam*

Scan the QR code to the right or go to www.electricalpereview.com and click "*START NOW FOR FREE*"

100% Free Trial:

Learn more about our online 11 week live class for the *NCEES® Power PE Exam with over 60 hours of live instruction*

Scan the QR code to the right or go to www.electricalpereview.com and click "*LIVE CLASS*"

Live Class:

More by Zach Stone, P.E. and Electrical PE Review, INC.

Study with power PE practice problems worked out on video FREE on our YouTube channel

Scan the QR code to the right or go to www.YouTube.com and search for "*Electrical PE Review*"

YouTube Channel:

Watch Interviews with our former students and learn the strategies they used to PASS the CBT format of the Power PE Exam

Scan the QR code to the right or go to www.YouTube.com and search for "*How to PASS the NCEES® Power PE Exam CBT*"

Student Interviews:

Have a question about a problem in this practice exam? Want to know more about our online class for the PE Exam? Email me!

Scan the QR code to the right or email me at zach@electricalpereview.com

Email Zach Stone, PE:

More by Zach Stone, P.E. and Electrical PE Review, INC.

Add me on LinkedIn:

Expand your career network by adding me as a contact on LinkedIn

Scan the QR code to the right or go to www.LinkedIn.com and search for "*Zach Stone, PE*"

Review This Book!

Learn something new?

Feel more confident about the subjects that appear on the PE exam?

Enjoy the detailed solutions in this book compared to other practice exams?

....Did you pass the PE exam?

Hundreds and hundreds of hours went into preparing this practice exam to help you, the person holding this book in their hands, prepare for and pass the NCEES® Power PE Exam.

I would greatly appreciate it if you could take one moment to leave an honest review for this book on Amazon.com.

Amazon reviews for this book help other engineers decide if they should purchase this book to help prepare for the PE exam, just like you did.

If I was able to help you in even a small way prepare for the PE exam by following along with the problems and solutions in this book, **please help me by taking a moment to leave an honest review on Amazon**. You can do it directly from your smartphone by scanning the QR code below with your camera app:

Thanks for taking the time to leave a review for this book!

Zach Stone, P.E.
Instructor
Electrical PE Review, INC

Review This Book:

Amazon.com

Made in United States
Troutdale, OR
10/18/2024